Digital System Design with VHDL

We work with leading authors to develop the
strongest educational materials in engineering,
bringing cutting-edge thinking and best learning
practice to a global market.

Under a range of well-known imprints, including
Prentice Hall, we craft high quality print and
electronic publications which help readers to
understand and apply their content, whether
studying or at work.

To find out more about the complete range of our
publishing, please visit us on the World Wide Web
at: www.pearsoned.co.uk

Digital System Design with VHDL

Second edition

Mark Zwoliński

Prentice Hall

An imprint of **Pearson Education**

Harlow, England · London · New York · Reading, Massachusetts · San Francisco · Toronto · Don Mills, Ontario · Sydney
Tokyo · Singapore · Hong Kong · Seoul · Taipei · Cape Town · Madrid · Mexico City · Amsterdam · Munich · Paris · Milan

To Kate, who had to listen to me shouting at the computer

Pearson Education Limited
Edinburgh Gate
Harlow
Essex CM20 2JE
England

and Associated Companies throughout the world

Visit us on the World Wide Web at:
www.pearsoned.co.uk

First published 2000
Second edition published 2004

© Pearson Education Limited 2000, 2004

ISBN 978-0-13-039985-4

British Library Cataloguing-in-Publication Data
A catalogue record for this book is available from the British Library

10 9 8 7 6
09 08 07

Typeset in 10/12 pt Times by 68

Printed and bound by Henry Ling Limited, at the Dorset Press, Dorchester, DT1 1HD

The publisher's policy is to use paper manufactured from sustainable forests.

Contents

Preface

....

About this book

When the first edition of this book was published, the idea of combining a text on digital design with one on VHDL seemed novel. At about the same time, several other books with similar themes were published. This book has now been adopted by several universities as a core text. Moreover, the first edition has now been translated into Polish and a low-cost edition has been produced for the People's Republic of China. This success and the competition convinced me that the idea had been good, but I was not convinced that I had achieved perfection the first time. This new edition corrects what I now feel to have been a mistake in the first version and adds two important topics. These changes are described later in this preface.

This book is intended as a student textbook for both undergraduate and postgraduate students. The majority of VHDL books are aimed at practising engineers. Therefore, some features of VHDL are not described at all in this book. Equally, aspects of digital design are covered that would not be included in a typical VHDL book.

Syllabuses for electrical, electronic and computer engineering degrees vary between countries and between universities or colleges. The material in this book has been developed over a number of years for second- and third-year undergraduates and for postgraduate students. It is assumed that students will be familiar with the principles of Boolean algebra and with combinational logic design. At the University of Southampton, the first-year undergraduate syllabus also includes introductions to synchronous sequential design and to programmable logic. This book therefore builds upon these foundations. It has often been assumed that topics such as VHDL are too specialized for second-year teaching and are best left to final-year or postgraduate courses. There are several good reasons why VHDL should be introduced earlier into the curriculum. With increasing integrated circuit complexity, there is a need from industry for graduates with knowledge of VHDL and the associated design tools. If left to the final year, there is little or no time for the student to apply such knowledge in project work. Second, conversations with colleagues from many countries suggest that today's students are opting for computer science or computer engineering courses in preference to electrical or electronic engineering. VHDL offers a means to interest computing-oriented students in hardware design. Finally, simulation and synthesis tools are now mature and available relatively cheaply to educational establishments on PC platforms.

Changes in the second edition

With hindsight, my mistake was to use `std_ulogic` instead of `std_logic` in most of the examples. From a purely educational point of view, the decision was correct as such usage would clearly indicate in simulation when outputs of blocks had been inadvertently joined together. From a practical point of view, this usage is at odds with most industrial practice and can cause problems with some EDA (electronic design automation) tools. All the examples have been revised to use `std_logic`. Several of the examples have also been simplified (e.g. by using direct instantiation). At the time of the first edition, there were some EDA tools that only supported the 1987 standard of VHDL. These have largely disappeared and therefore I have tended to use constructs from the newer 1993 standard in preference in this edition. There has also been a 2002 revision to the standard. Although there are almost no tools that support the new standard at the time of writing, the changes are minimal and the only significant change (the form of shared variables) has been extensively discussed in Appendix C. I was also swimming against the tide in insisting on the use of the IEEE `numeric_std` package (as opposed to `std_logic_arith`), but I think I have been proved correct in that choice.

The two major additions take two forms. First, several chapters now include sections on writing testbenches. The verification of VHDL models by simulation is critical to producing correct hardware. It is reasonable to estimate that at least half of all VHDL written is in the form of testbenches for verifying models. Because this aspect is so important, the material has been included in the chapters where it is needed, not in a single chapter on testbench design. I would strongly encourage the reader to simulate the models in the text and to use the testbench examples to assist in this.

The second addition is a new chapter on VHDL-AMS and mixed-signal modelling. All digital hardware has to interact with the 'real' world at some point. Although mixed-signal simulators have been available for over 15 years, their use has been limited by the difficulty in writing interface models between the digital and analogue domains. VHDL-AMS integrates the two worlds and several mixed-signal simulators are now available. This chapter is not intended to be a comprehensive tutorial on converter design, nor on all the details of VHDL-AMS, but I hope it will encourage designers to attempt to model their systems as a whole.

Structure of this book

Chapter 1 introduces the ideas behind this book, namely the use of electronic design automation tools and CMOS and programmable logic technology. We also consider some engineering problems, such as noise margins and fan-out. In Chapter 2, the principles of Boolean algebra and of combinational logic design are reviewed. The important matter of timing and the associated problem of hazards are discussed. Some basic techniques for representing data are discussed.

VHDL is introduced in Chapter 3 through basic logic gate models. The importance of documented code is emphasized. We show how to construct netlists of basic gates and how to model delays through gates. We also discuss parameterized models and constant and unconnected inputs and outputs. The idea of using VHDL to verify VHDL models by using testbenches is introduced. Finally, we briefly introduce the concept of configurations.

In Chapter 4, a variety of modelling techniques are described. Combinational building blocks, buffers, decoders, encoders, multiplexers, adders and parity checkers are modelled using a range of concurrent and sequential VHDL coding constructs. The VHDL models of hardware introduced in this chapter and in Chapters 5, 6 and 7 are, in principle, synthesizable, although discussion of exactly what is supported is deferred until Chapter 9. Testbench design styles are again discussed here. In addition, the IEEE dependency notation is introduced.

Chapter 5 is probably the most important chapter of the book and discusses what might be considered the cornerstone of digital design: the design of finite state machines. The ASM chart notation is used. The design process from ASM chart to D flip-flops and next state and output logic is described. VHDL models of state machines are introduced.

Chapter 6 introduces various sequential building blocks: latches, flip-flops, registers, counters, memory and a sequential multiplier. The same style as in Chapter 4 is used, with IEEE dependency notation, testbench design and the introduction of VHDL coding constructs.

In Chapter 7 the concepts of the previous three chapters are combined. The ASM chart notation is extended to include coupled state machines and registered outputs, and hence to datapath-controller partitioning. From this, we explain the idea of instructions in hardware terms and go on to model a very basic microprocessor in VHDL. This provides a vehicle to introduce VHDL subroutines and packages.

VHDL remains primarily a modelling language. Chapter 8 describes the operation of a VHDL simulator. The idea of event-driven simulation is first explained and the specific features of a VHDL simulator are then discussed. Although the entire VHDL language can be simulated, some constructs simulate more efficiently than others; therefore techniques for writing models that are more efficient are discussed. File operations are also discussed in this chapter because such functionality is only appropriate to simulation models.

The other, increasingly important, role of VHDL is as a language for describing synthesis models, as discussed in Chapter 9. The dominant type of synthesis tool available today is for RTL synthesis. Such tools can infer the existence of flip-flops and latches from a VHDL model. These constructs are described. Conversely, flip-flops can be created in error if the description is poorly written, and common pitfalls are described. The synthesis process can be controlled by constraints. Because these constraints are outside the language, they are discussed in general terms. Suitable constructs for FPGA synthesis are discussed. Finally, behavioural synthesis, which promises to become an important design technology, is briefly examined.

Chapters 10 and 11 are devoted to the topics of testing and design for test. This area has often been neglected, but is now recognized as being an important part of the design process. In Chapter 10 the idea of fault modelling is introduced. This is followed by test generation methods. The efficacy of a test can be determined by fault simulation. At the time of writing, there are no commercial VHDL-based fault simulators available. The final section of this chapter shows how fault modelling and fault simulation can be performed using a standard VHDL simulator. The VHDL code also introduces constructs such as pointers and shared (global) variables.

In Chapter 11, three important design-for-test principles are described: scan path, built-in self-test (BIST) and boundary scan. This has always been a very dry subject, but a VHDL simulator can be used, for example, to show how a BIST structure can generate different signatures for fault-free and faulty circuits. Boundary scan uses a subset of VHDL to describe the test structures used on a chip, and an example is given.

We use VHDL as a tool for exploring anomalous behaviour in asynchronous sequential circuits in Chapter 12. Although the predominant design style is currently synchronous, it is likely that digital systems will increasingly consist of synchronous circuits communicating asynchronously with each other. We introduce the concept of the fundamental mode and show how to analyze and design asynchronous circuits. We use VHDL simulations to illustrate the problems of hazards, races and setup and hold time violations. We also discuss the problem of metastability.

The final chapter introduces VHDL-AMS and mixed-signal modelling. Brief descriptions of Digital to Analogue Converters (DACs) and Analogue to Digital Converters (ADCs) are given. VHDL-AMS constructs to model such converters are given. We also introduce the idea of a Phase-Locked Loop (PLL) here and give a simple mixed-signal model.

Three appendices are included. The first appendix lists the various VHDL-related standards and speculates on the future development of VHDL. The second appendix briefly describes the Verilog hardware description language. Verilog is the major alternative to VHDL and it is likely that designers will have to be familiar with both. The third appendix covers shared variables, in particular the differences between the 1993 and 2002 versions.

At the end of each chapter a number of exercises have been included. These exercises are almost secondary to the implicit instruction in each chapter to simulate and, where appropriate, synthesize each VHDL example. To perform these simulation and synthesis tasks, the reader may have to write his or her own testbenches and constraints files. The examples are available on the World Wide Web at the address given in the next section.

How to use this book

Obviously, this book can be used in a number of different ways, depending on the level of the course. At the University of Southampton, I am using the material as follows.

Second year of MEng/BEng in Electronic Engineering and Computer Engineering

Chapters 1 and 2 are review material, which the students would be expected to read independently. Lectures then cover the material of Chapters 3, 4, 5, 6 and 7. Some of this material can be considered optional, such as Sections 3.8, 6.3 and 6.7.

Additionally, constructs such as `with select` could be omitted if time presses. The single-stuck fault model of Section 10.2 and the principles of test pattern generation in Section 10.3, together with the principles of scan design in Section 11.2, would also be covered in lectures.

Third year of MEng/BEng in Electronic Engineering and Computer Engineering
Students would be expected to independently re-read Chapters 3 to 7. Lectures would cover Chapters 8, 9, 10, 11 and 12. VHDL-AMS, Chapter 13, is currently covered in a fourth-year module.

In both years, students need to have access to a VHDL simulator and an RTL synthesis tool in order to use the examples in the text. In the second year, a group design exercise involving synthesis to an FPGA would be an excellent supplement to the material. In the third year at Southampton, all students do an individual project. There is no additional formal laboratory work. Some of the individual projects will involve the use of VHDL.

Mark Zwoliński
Southampton
May 2003

Web resources

A website accompanies Digital System Design with VHDL by Mark Zwoliński. Visit the site at www.booksites.net/Zwolinski. Here you will find valuable teaching and learning material including all the VHDL examples from the text (differentiated between VHDL'87 and VHDL'93 versions), and links to sites with VHDL tools.

Acknowledgements

I would like to thank all those who pointed out errors in the first edition of this book, in particular Egbert Molenkamp of the University of Twente, the Netherlands, and Marian Adamski, Marek Węgrzyn and Zbigniew Skrowroński at the University of Zielona Góra, Poland.

Some of the material in Chapter 13 was produced in collaboration with Doulos Ltd.

The publishers are grateful to the following for permission to reproduce material:

Figure 1.11 is reproduced with the permission of Lattice Semiconductor Corporation; Figure 1.15 copyright ©1999 Xilinx, Inc. All rights reserved, XC4000E and XC4000X Series Field Programmable Gate Arrays.

In some instances we have been unable to trace the owners of copyright material, and we would appreciate any information that would enable us to do so.

Finally, I would like to thank several cohorts of students to whom I have delivered this material and whose comments have encouraged me to think about better ways of explaining these ideas.

Chapter 1

Introduction

In this chapter we will review the design process, with particular emphasis on the design of digital systems using hardware description languages such as VHDL. The technology of CMOS integrated circuits will be briefly revised and programmable logic technologies will be discussed. Finally, the relevant electrical properties of CMOS and programmable logic are reviewed.

1.1 Modern digital design

Electronic circuit design has traditionally fallen into two main areas: analogue and digital. These subjects are usually taught separately, and electronics engineers tend to specialize in one area. Within these two groupings there are further specializations, such as radio frequency analogue design, digital integrated circuit design, and, where the two domains meet, mixed-signal design. In addition, of course, software engineering plays an increasingly important role in embedded systems.

Digital electronics is ever more significant in consumer goods. Cars have sophisticated control systems. Many homes now have personal computers. Products that used to be thought of as analogue, such as radio, television and telephones, are or are becoming digital. Digital compact discs have almost entirely replaced analogue LPs for recorded audio. With these changes, the lifetimes of products have lessened. In a period of less than a year, new models will probably have replaced all the digital electronic products in your local store.

1.1.1   Design automation

To keep pace with this rapid change, electronics products have to be designed extremely quickly. Analogue design is still a specialized (and well-paid) profession. Digital design has become very dependent on computer-aided design (CAD) – also known as design automation (DA) or electronic design automation (EDA). The EDA tools allow two tasks to be performed: *synthesis*, in other words the translation of a specification into an actual implementation of the design; and *simulation*, in which the specification or the detailed implementation can be exercised in order to verify correct operation.

Synthesis and simulation EDA tools require that the design be transferred from the designer's imagination into the tools themselves. This can be done by drawing a diagram of the design using a graphical package. This is known as *schematic capture*. Alternatively, the design can be represented in a textual form, much like a software program. Textual descriptions of digital hardware can be written in a modified programming language, such as C, or in a *hardware description language* (HDL). Over the past 30 years or so, a number of HDLs have been designed. Two HDLs are in common usage today: Verilog and VHDL (VHSIC Hardware Description Language, where VHSIC stands for Very High Speed Integrated Circuit). Standard HDLs are important because they can be used by different CAD tools from different tool vendors. In the days before Verilog and VHDL, every tool had its own HDL, requiring laborious translation between HDLs, for example to verify the output from a synthesis tool with another vendor's simulator.

1.1.2   Logic gates

The basic building blocks of digital circuits are *gates*. A gate is an electronic component with a number of inputs and, generally, a single output. The inputs and the outputs are normally in one of two states: logic 0 or logic 1. These logic values are represented by voltages (for instance, 0 V for logic 0 and 3.3 V for logic 1) or currents. The gate itself performs a logical operation using all of its inputs to generate the output. Ultimately, of course, digital gates are really analogue components, but for simplicity we tend to ignore their analogue nature.

It is possible to buy a single integrated circuit containing, say, four identical gates, as shown in Figure 1.1. (Note that two of the connections are for the positive and negative power supplies to the device. These connections are not normally shown in logic diagrams.) A digital system could be built by connecting hundreds of such devices together – indeed many systems have been designed in that way. Although the individual integrated circuits might cost as little as 10 cents each, the cost of designing the printed circuit board for such a system and the cost of assembling the board are very significant and this design style is no longer cost-effective.

Much more complicated functions are available as mass-produced integrated circuits, ranging from flip-flops through to microprocessors. With increasing complexity comes flexibility – a microprocessor can be programmed to perform a near-infinite variety of tasks. Digital system design therefore consists, in part, of taking standard components and connecting them together. Inevitably, however, some aspect of the functionality will not be available as a standard device. The designer is then left with

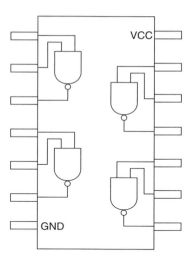

Figure 1.1   Small-scale integrated circuit.

the choice of implementing this functionality from discrete gates or of designing a specialized integrated circuit to perform that task. While this latter task may appear daunting, it should be remembered that the cost of a system will depend to a great extent not on the cost of the individual components but on the cost of connecting those components together.

1.1.3   ASICs and FPGAs

The design of a high-performance, full-custom integrated circuit (IC) is, of course, a difficult task. In full-custom IC design, *everything*, down to and including individual transistors, may be designed (although libraries of parts are, of course, used). For many years, however, it has been possible to build semi-custom integrated circuits using *gate arrays*. A gate array, as its name suggests, is an integrated circuit on which an array of logic gates has been created. The design of an *application-specific integrated circuit* (ASIC) using a gate array therefore involves the definition of how the gates in the array should be connected. In practical terms, this means that one or two layers of metal interconnect must be designed. Since an integrated circuit requires seven or more processing stages, all the processing steps other than the final metallization can be completed in advance. Because the uncommitted gate arrays can be produced in volume, the cost of each device is relatively small.

The term ASIC is often applied to full-custom and semi-custom integrated circuits. Another class of integrated circuit is that of *programmable logic*. The earliest programmable logic devices (PLDs) were *programmable logic arrays* (PLAs). Like gate arrays, these consist of arrays of uncommitted logic, but unlike *mask-programmable* gate arrays, the configuration of the array is determined by applying a large (usually negative) voltage to individual connections. The general structure of a PLA is shown in Figure 1.2. The PLA has a number of inputs (A, B, C) and outputs (X, Y, Z),

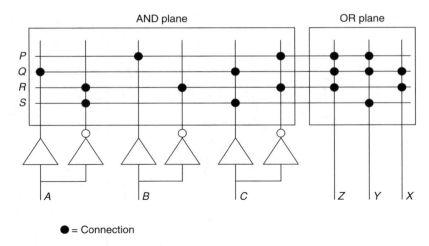

Figure 1.2 PLA structure.

an AND-plane and an OR-plane. Connections between the inputs and the product terms (P, Q, R, S) and between the product terms and outputs are shown; the remaining connections have been removed as part of the programming procedure. Some PLAs may be reprogrammed electrically or by restoring the connections by exposing the device to ultraviolet light. PALs (programmable array logic) extend the idea of PLAs to include up to 12 flip-flops. In recent years, programmable devices have become much more complex and include CPLDs (complex PLDs) and FPGAs (*field programmable gate arrays*). FPGAs are described in more detail in Section 1.3.

1.1.4 Design flow

Most digital systems are sequential, that is they have states, and the outputs depend on the present state. Some early designs of computer were *asynchronous*; in other words, the transition to a new state happened as soon as inputs had stabilized. For many years, digital systems have tended to be *synchronous*. In a synchronous system, the change of state is triggered by one or more clock signals. In order to design reliable systems, formal design methodologies have been defined. The design of a (synchronous sequential) digital system using discrete gates would therefore proceed as follows.

1. Write a specification.
2. If necessary, partition the design into smaller parts and write a specification for each part.
3. From the specification draw a state machine chart. This shows each state of the system and the input conditions that cause a change of state, together with the outputs in each state.
4. Minimize the number of states. This is optional and may not be useful in all cases.
5. Assign Boolean variables to represent each state.

6. Derive the next state and output logic.

7. Optimize the next state and output logic to minimize the number of gates needed.

8. Choose a suitable placement for the gates in terms of which gates share integrated circuits and in terms of where each integrated circuit is placed on the printed circuit board.

9. Design the routing between the integrated circuits.

In general, steps 1 and 2 cannot be avoided. This is where the creativity of the designer is needed. Most books on digital design concentrate on steps 3 to 7. Steps 8 and 9 can be performed manually, but placement and routing was one of the first tasks to be successfully automated. It is possible to simulate the design at different stages if it is converted into a computer-readable form. Typically, in order to perform the placement and routing, a schematic capture program would be used at around step 7, such that the gate-level structure of the circuit would be entered. This schematic could be converted to a form suitable for a logic simulator. After step 9 had been completed, the structure of the circuit, including any delays generated by the resistance and capacitance of the interconnect, could be extracted and again simulated.

The implementation of digital designs on ASICs or FPGAs therefore involves the configuration of connections between predefined logic blocks. As noted, we cannot avoid steps 1 and 2 above, and steps 8 and 9 can be done automatically. The use of an HDL, which in the case of this book is VHDL, means that the design can be entered into a CAD system and simulated at step 3 or 4, rather than step 7. So-called register transfer level (RTL) synthesis tools automate steps 6 and 7. Step 4 still has to be done by hand. Step 5 can be automated, but now the consequences of a particular state assignment can be assessed very quickly. Behavioural synthesis tools are starting to appear that automate the process from about step 2 onwards. Figure 1.3 shows the overall design flow for RTL synthesis-based design.

Because of this use of EDA tools to design ASICs and FPGAs, a book such as this can concentrate on higher-level aspects of design, in particular the description of functional blocks in an HDL. Many books on digital design describe multiple output and multi-level logic minimization, including techniques such as the Quine–McCluskey algorithm. Here, we assume that a designer may occasionally wish to minimize expressions with a few variables and a single output, but if a complex piece of combinational logic is to be designed a suitable EDA tool is available that will perform the task quickly and reliably.

1.2 CMOS technology

As noted, even digital gates can be thought of as analogue circuits. The design of individual gates is therefore a circuit design problem. Hence there exist a wide variety of possible circuit structures. Very early digital computers were built using vacuum tubes. These gave way to transistor circuits in the 1960s and 1970s. There are two major types of transistor: bipolar junction transistors (BJTs) and field effect transistors (FETs). Logic families such as TTL (transistor–transistor logic) and ECL (emitter–collector logic) use BJTs. Today, the dominant (but not exclusive) technology is CMOS, which uses FETs. CMOS derives its

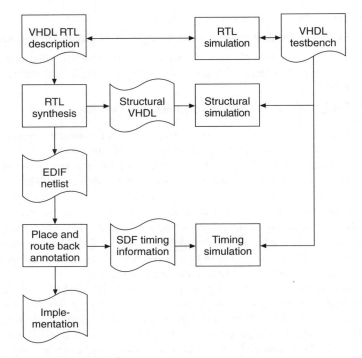

Figure 1.3   RTL synthesis design flow.

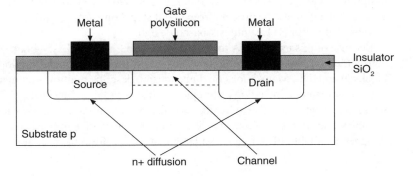

Figure 1.4   NMOS transistor structure.

name from the particular type of FET used – the MOSFET (metal oxide semiconductor FET). CMOS therefore stands for complementary MOS, as two types of MOS device are used. MOS is, in fact, a misnomer; a better term is IGFET (insulated gate FET).

The structure of an n-type (NMOS) MOS transistor is shown in Figure 1.4, which is not drawn to scale. The substrate is the silicon wafer that has been doped to make it p-type. The thickness of the substrate is therefore significantly greater than the other transistor dimensions. Two heavily doped regions of n-type silicon are created for each transistor. These form the *source* and *drain*. In fact, the source and drain are inter-changeable, but by convention the drain–source voltage is usually positive. Metal connections are made to the source and drain. The polycrystalline silicon (polysilicon) gate is separated from the rest of the device by a layer of silicon dioxide insulator. Originally the gate would have been metal – hence the name MOS was derived from the structure of the device (metal oxide semiconductor).

When the gate voltage is the same as the source voltage, the drain is insulated from the source. As the gate voltage rises, the gate–oxide–semiconductor sandwich acts as a capacitor, and negative charge builds up on the surface of the semiconductor. At a critical *threshold voltage* the charge is sufficient to create a channel of n-type silicon between the source and drain. This acts as a conductor between the source and the drain. Therefore the NMOS transistor can be used as a switch that is open when the gate voltage is low and closed when the gate voltage is high.

A PMOS transistor is formed by creating heavily doped p-type drain and source regions in an n-type substrate. A PMOS transistor conducts when the gate voltage is low and does not conduct when the gate voltage is high.

Symbols for NMOS transistors are shown in Figures 1.5(a) and (b). The substrate is also known as the *bulk*, hence the symbol B. In digital circuits, the substrate of NMOS transistors is always connected to ground (logic 0) and hence can be omitted from the symbol, as shown in Figure 1.5(b). Symbols for PMOS transistors are shown in Figures 1.5(c) and (d). Again the bulk connection is not shown in Figure 1.5(d), because in digital circuits the substrate of a PMOS transistor is always connected to the positive supply voltage (logic 1).

A logical inverter (a NOT gate) can be made from an NMOS transistor and a resistor, or from a PMOS transistor and a resistor, as shown in Figures 1.6(a) and (b), respectively. VDD is the positive supply voltage (3.3 V to 5 V); GND is the ground connection (0 V). The resistors have a reasonably high resistance, say 10 kΩ. When IN is at logic 1 (equal to the VDD voltage), the NMOS transistor in Figure 1.6(a) acts as a closed switch.

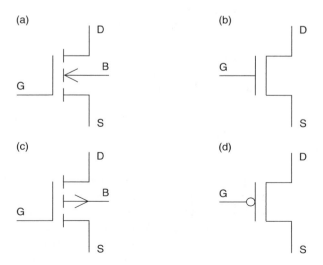

Figure 1.5 MOS transistor symbols: (a), (b) NMOS; (c), (d) PMOS.

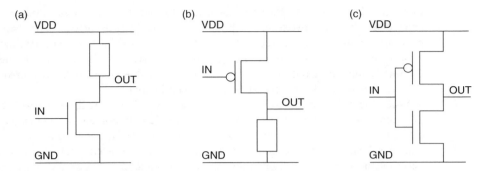

Figure 1.6   MOS inverters: (a) NMOS; (b) PMOS; (c) CMOS.

Because the resistance of the NMOS transistor, when it is conducting, is much less than that of the resistor, OUT is connected to GND, giving a logic 0 at that node. In the circuit of Figure 1.6(b), a logic 1 at IN causes the PMOS transistor to act as an open switch. The resistance of the PMOS transistor is now much greater than that of the resistance, so OUT is connected to GND via the resistor. Again a logic 0 is asserted at OUT.

A logic 0 at IN causes the opposite effects. The NMOS transistor becomes an open switch, causing OUT to be connected to VDD by the resistor; the PMOS transistor becomes a closed switch with a lower resistance than the resistor and again OUT is connected to VDD.

Figure 1.6(c) shows a CMOS inverter. Here, both PMOS and NMOS transistors are used. A logic 1 at IN will cause the NMOS transistor to act as a closed switch and the PMOS transistor to act as an open switch, giving a 0 at OUT. A logic 0 will have the opposite effect: the NMOS transistor will be open and the PMOS transistor will be closed. The name CMOS comes from complementary MOS – the NMOS and PMOS transistors complement each other.

Current flows in a semiconductor as electrons move through the crystal matrix. In p-type semiconductors it is convenient to think of the charge being carried by the absence of an electron, a 'hole'. The mobility of holes is less than that of electrons (i.e. holes move more slowly through the crystal matrix than electrons). The effect of this is that the gain of a PMOS transistor is less than that of the same-sized NMOS transistor. Thus to build a CMOS inverter with symmetrical characteristics, in the sense that a 0 to 1 transition happens at the same rate as a 1 to 0 transition, requires that the gain of the PMOS and NMOS transistors be made the same. This is done by varying the widths of the transistors (assuming the lengths are the same) such that the PMOS transistor is about 2.5 times as wide as the NMOS transistor. As will be seen, this effect is compensated for in CMOS NAND gates, where similarly sized NMOS and PMOS transistors can be used. CMOS NOR gates, however, do require the PMOS transistors to be scaled. Hence, NAND gate logic is often preferred for CMOS design.

Two-input CMOS NAND and NOR gates are shown in Figures 1.7(a) and (b), respectively. The same reasoning as used in the description of the inverter may be applied. A logic 1 causes an NMOS transistor to conduct and a PMOS transistor to be open; a logic 0 causes the opposite effect. NAND and NOR gates with three or more

(a)

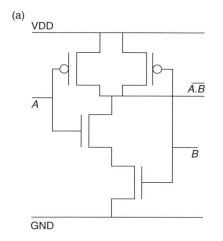

(b)

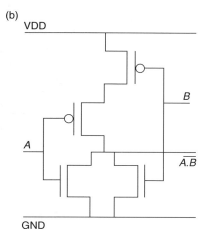

Figure 1.7 (a) CMOS NAND; (b) CMOS NOR.

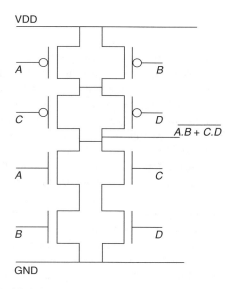

Figure 1.8 CMOS AND–OR–Invert.

inputs can be constructed using similar structures. Note that in a NAND gate all the PMOS transistors must have a logic 0 at their gates for the output to go high. As the transistors are working in parallel, the effect of the lower mobility of holes on the gain of the transistors is overcome.

Figure 1.8 shows a CMOS AND–OR–Invert structure. The function $\overline{(A.B) + (C.D)}$ can be implemented using eight transistors compared with the 14 needed for three NAND/NOR gates and an inverter.

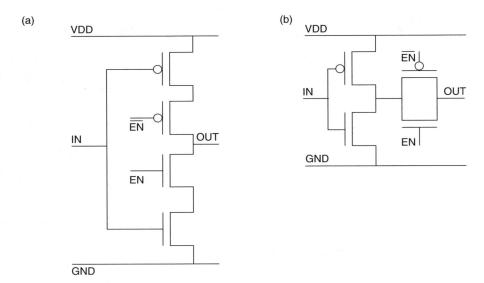

Figure 1.9   CMOS three-state buffer.

A somewhat different type of structure is shown in Figure 1.9(a). This circuit is a three-state buffer. When the *EN* input is at logic 1, and the $\overline{EN}$ input is at logic 0, the two inner transistors are conducting and the gate inverts the *IN* input as normal. When the *EN* input is at logic 0 and the $\overline{EN}$ input is at logic 1, neither of the two inner transistors is conducting and the output floats. The $\overline{EN}$ input is derived from *EN* using a standard CMOS inverter. An alternative implementation of a three-state buffer is shown in Figure 1.9(b). Here a transmission gate follows the CMOS inverter. The NMOS and PMOS transistors of the transmission gate are controlled by complementary signals. When *EN* is at logic 1 and $\overline{EN}$ is at logic 0, both transistors conduct; otherwise both transistors are open circuit.

Figure 1.10(a) shows a two-input multiplexer constructed from transmission gates while Figures 1.10(b) and (c) show an exclusive OR gate and a D latch, respectively, that both use CMOS transmission gates. All these circuits use fewer transistors than the equivalent circuits constructed from standard logic gates. It should be noted, however, that the simulation of transmission gate circuits can be problematic. VHDL, in particular, is not well suited to this type of transistor-level modelling, and we do not give any examples in this book, other than of general three-state buffers.

1.3   Programmable logic

While CMOS is currently the dominant technology for integrated circuits, for reasons of cost and performance, many designs can be implemented using *programmable logic*. The major advantage of programmable logic is the speed of implementation. A programmable logic device can be configured on a desktop in seconds, or at most minutes.

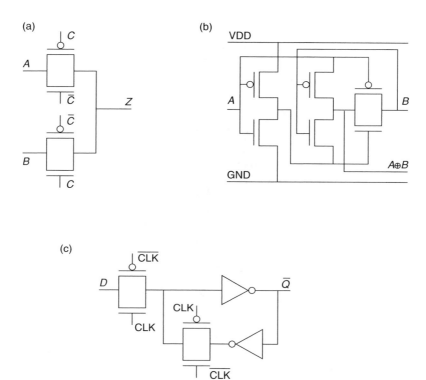

Figure 1.10 CMOS transmission gate circuits: (a) multiplexer; (b) XOR; (c) D latch.

The fabrication of an integrated circuit can take several weeks. The cost per device of a circuit built in programmable logic may be greater than that of a custom integrated circuit, and the performance, in terms of both speed and functionality, is likely to be less impressive than that of CMOS. These apparent disadvantages are often outweighed by the ability to rapidly produce working integrated circuits. Thus programmable logic is suited to prototypes, but also increasingly to small production volumes.

One recent application of programmable devices is as *reconfigurable logic*. A system may perform different functions at different points in time. Instead of having all the functionality available all the time, one piece of hardware may be reconfigured to implement the different functions. New functions, or perhaps better versions of existing functions, could be downloaded from the Internet. Such applications are likely to become more common in future.

There are a number of different technologies used for programmable logic by different manufacturers. The simplest devices, *programmable logic arrays* (PLAs), consist of two programmable planes, as shown in Figure 1.2. In reality, both planes implement a NOR function. The device is programmed by breaking connections. Most simple programmable devices use some form of floating gate technology. Each connection in the programmable planes consists of a MOS transistor. This transistor has two gates – one is connected to the input, while the second, between the first gate

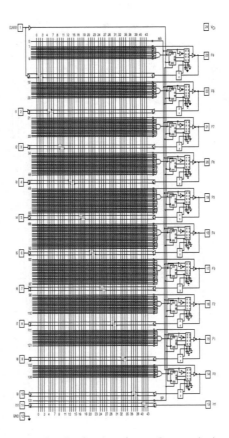

Figure 1.11   PAL structure (Lattice Semiconductor Corporation).

and the channel, floats. When the appropriate negative voltage is applied to the device, the floating gate can have a large charge induced on it. This charge will exist indefinitely. If the charge exists on the floating gate, the device is disabled; if the charge is not there, the device acts as a normal transistor. The mechanisms for putting the charge on the device include *avalanche* or *hot electron injection* (EPROM) and *Fowler–Nordheim tunnelling* (EEPROM and Flash devices). These devices can be reprogrammed electrically.

PALs have a programmable AND plane and a fixed OR plane, and usually include registers, as shown in Figure 1.11. More complex PLDs (CPLDs) consist effectively of a number of PAL-like macrocells that can communicate through programmable interconnect, as shown in Figure 1.12.

More complex still are *field programmable gate arrays* (FPGAs). FPGAs have a different type of architecture from CPLDs and are implemented in different technologies. Each FPGA vendor tends to have its own architecture – we will discuss two particular architectures here. Actel FPGAs consist of an array of combinational and sequential cells as shown in Figure 1.13. The combinational and sequential cells are shown in

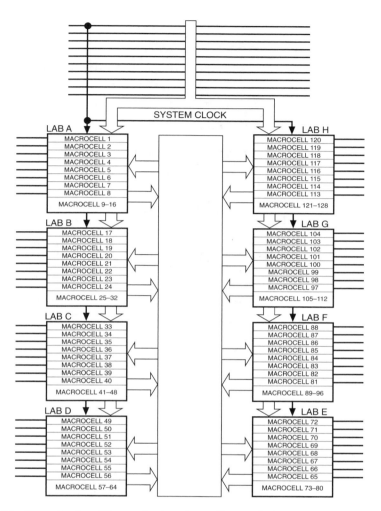

Figure 1.12   CPLD structure (Cypress Semiconductor Corporation).

Figures 1.14(a) and (b), respectively. Actel FPGAs are configured using an antifuse technology. In other words, a connection is normally open circuit, but the application of a suitably large voltage causes a short-circuit to be formed. This configuration is not reversible, unlike EPROM or Flash technology. Once made, a short-circuit has a resistance of around 50 Ω, which limits the fan-out, as described below.

Xilinx FPGAs are implemented in static RAM technology. Unlike most programmable logic, the configuration is therefore volatile and must be restored each time power is applied to the circuit. Again, these FPGAs consist of arrays of logic cells. One such cell is shown in Figure 1.15. Each of these cells can be programmed to implement a range of combinational and sequential functions. In addition to these logic cells, there exists programmable interconnect, including three-state buffers.

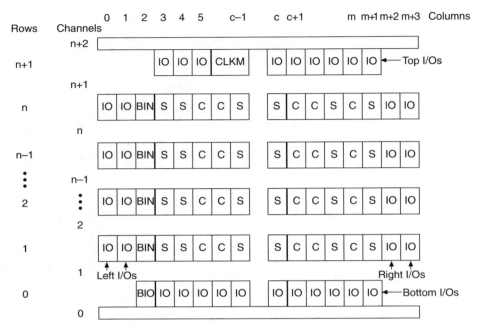

Figure 1.13 Actel FPGA (Actel Corporation).

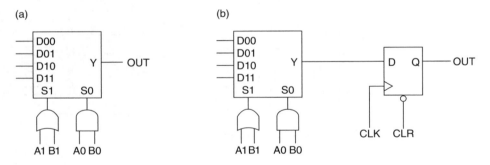

Figure 1.14 Actel FPGA cells: (a) combinational; (b) sequential (Actel Corporation).

1.4 Electrical properties

1.4.1 Noise margins

Although it is common to speak of a logic 1 being, say, 2.5 V and a logic 0 being 0 V, in practice a range of voltages represent a logic state. A range of voltages may be recognized as a logic 1, and similarly one voltage from a particular range may be generated for a logic 1. Thus we can describe the logic states in terms of the voltages shown in Table 1.1.

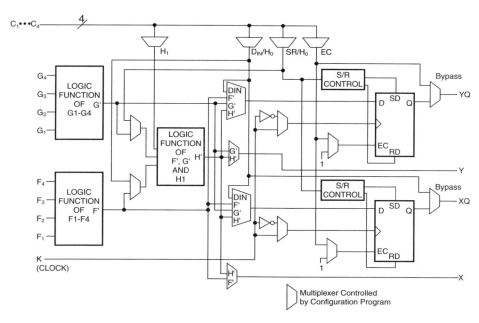

Figure 1.15   Xilinx FPGA logic cell (Xilinx, Inc.).

Table 1.1	Typical voltage levels for CMOS circuits with a supply voltage of 2.5 V.	
Parameter	Description	Typical CMOS value
V_{IHmax}	Maximum voltage recognized as a logic 1	2.5 V
V_{IHmin}	Minimum voltage recognized as a logic 1	1.35 V
V_{ILmax}	Maximum voltage recognized as a logic 0	1.05 V
V_{ILmin}	Minimum voltage recognized as a logic 0	0.0 V
V_{OHmax}	Maximum voltage generated as a logic 1	2.5 V
V_{OHmin}	Minimum voltage generated as a logic 1	1.75 V
V_{OLmax}	Maximum voltage generated as a logic 0	0.75 V
V_{OLmin}	Minimum voltage generated as a logic 0	0.0 V

The transfer characteristic for a CMOS inverter is illustrated in Figure 1.16. The *noise margin* specifies how much noise, from electrical interference, can be added to a signal before a logic value is misinterpreted. From Table 1.1, it can be seen that the maximum voltage that a gate will generate to represent a logic 0 is 0.75 V. Any voltage up to 1.05 V is, however, recognized as a logic 0. Therefore there is a 'spare' 0.3 V, and any noise added to a logic 0 within this band will be accepted. Similarly, the difference

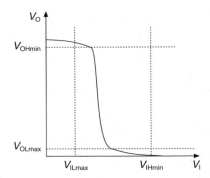

Figure 1.16 Transfer characteristic of CMOS inverter.

between the minimum logic 1 voltage generated and the minimum recognized is 0.4 V. The noise margins are calculated as:

$$\text{NM}_\text{L} = V_\text{ILmax} - V_\text{OLmax}$$
$$\text{NM}_\text{H} = V_\text{OHmin} - V_\text{IHmin}$$

In general, the bigger the noise margin, the better.

1.4.2 Fan-out

The fan-out of a gate is the number of other gates that it can drive. Depending on the technology, there are two ways to calculate the fan-out. If the input to a gate is resistive, as is the case with TTL or antifuse technology, the fan-out is calculated as the ratio of the current that a gate can output to the amount of current required to switch the input of a gate. For example, 74ALS series gates have the input and output currents specified in Table 1.2.

Two fan-out figures can be calculated:

$$\text{Logic 1 fan-out} = \frac{I_\text{OHmax}}{I_\text{IHmax}} = \frac{400\ \mu\text{A}}{20\ \mu\text{A}} = 20$$
$$\text{Logic 0 fan-out} = \frac{I_\text{OLmax}}{I_\text{ILmax}} = \frac{8\ \text{mA}}{100\ \mu\text{A}} = 80$$

Obviously the smaller of the two figures must be used.

CMOS gates draw almost no DC input current because there is no DC path between the gate of a transistor and the drain, source or substrate of the transistor. Therefore it would appear that the fan-out of CMOS circuits is very large. A different effect applies in this case. Because the gate and substrate of a CMOS gate form a capacitor, it takes a finite time to charge that capacitor, and hence the fan-out is determined by how fast the circuit is required to switch. In addition, the interconnect between two gates has

Table 1.2 Input and output currents for 74ALS series TTL gates.

I_{IHmax}	Maximum logic 1 input current	20 μA
I_{ILmax}	Maximum logic 0 input current	−100 μA
I_{OHmax}	Maximum logic 1 output current	−400 μA
I_{OLmax}	Maximum logic 0 output current	8 mA

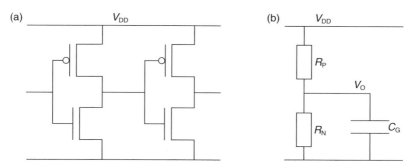

Figure 1.17 (a) CMOS inverter driving CMOS inverter; (b) equivalent circuit.

a capacitance. In high-performance circuits the effect of the interconnect can dominate that of the gates themselves. Obviously, the interconnect characteristics cannot be estimated until the final layout of the circuit has been completed.

Figure 1.17(a) shows one CMOS inverter driving another. Figure 1.17(b) shows the equivalent circuit. If the first inverter switches from a logic 1 to a logic 0 at $t = 0$, and if we assume that the resistance of NMOS transistor is significantly less than the resistance of the PMOS transistor, V_O is given by:

$$V_O = V_{DD}e^{-t/R_N C_G}$$

From Table 1.1 above, the minimum value of V_O that would be recognized as a logic 1 is 1.35 V and the maximum value of V_O that would be recognized as a logic 0 is 1.05 V. For example, if V_{DD} is 2.5 V, R_N is 100 Ω and C_G is 100 pF, we can see that the time taken for V_O to drop from 1.35 V to 1.05 V is given by:

$$t = -100 \times 100 \times 10^{-12} \times \ln\frac{1.05}{2.5} + 100 \times 100 \times 10^{-12} \times \ln\frac{1.35}{2.5}$$
$$= 2.5 \text{ ns}$$

If two inverters are driven, the capacitive load doubles, so the switching time doubles. Therefore, although a CMOS gate can drive an almost unlimited number of other gates at a fixed logic level, the fan-out is limited by the speed required of the circuit.

Summary

Digital design is no longer a matter of taking small-scale integrated circuits and connecting them together. Programmable logic devices are an important alternative to full-custom integrated circuits. A number of different technologies exist for PLDs. These different technologies impose different constraints on the designer.

Further reading

The best source of information about different families of programmable logic is the manufacturers themselves. The entire data books are now available on the Web. These generally include electrical information, design advice and hints for programming using VHDL. In general, it is easy to guess the Web addresses. For example, Xilinx are at http://www.xilinx.com/ and Actel are at http://www.actel.com/

Exercises

1.1 Find examples of the following components in a 74LS/74HC data book (or on the Web):

- 4-bit universal shift register
- 4-bit binary counter
- 8-bit priority encoder
- 4-bit binary adder
- 4-bit ALU

1.2 Find examples of PLDs, CPLDs and FPGAs from manufacturers' data books or from the Web. Compare the following factors:

- technologies
- performance
- cost
- programmability (e.g. use of VHDL)
- testability

1.3 How is VHDL used in the design process?

1.4 FPGAs are available in a number of sizes. Given that smaller FPGAs will be cheaper, what criteria would you use to estimate the required size of an FPGA, prior to detailed design?

1.5 A digital system may be implemented in a number of different technologies. List the main types available and comment on the advantages and disadvantages of each option. If you were asked to design a system with about 5000 gates and which was expected to sell about 10 000 units, which hardware option would you choose and why?

Chapter 2

Combinational logic design

Digital design is based on the processing of binary signals. In this chapter, we will review the principles of Boolean algebra and the minimization of Boolean expressions. Hazards and basic numbering systems will also be discussed.

2.1 Boolean algebra

2.1.1 Values

Digital design uses a two-value algebra. Signals can take one of two values that can be represented by

ON and OFF, or
TRUE and FALSE, or
1 and 0.

2.1.2 Operators

The algebra of two values, known as Boolean algebra after George Boole (1815–1864), has five basic operators. In decreasing order of precedence (i.e. in the absence of parentheses, operations at the top of the list should be evaluated first), these are:

- NOT
- AND
- OR
- IMPLIES
- EQUIVALENCE

The last two operators are not normally used in digital design. These operators can be used to form expressions, for example:

$$A = 1$$
$$B = C \text{ AND } 0$$
$$F = \overline{(A + B.C)}$$
$$Z = (\overline{A} + B).(A + \overline{B})$$

The symbol '+' means 'OR', '.' means 'AND', and the overbar, e.g. '$\overline{A}$', means 'NOT A'.

2.1.3   Truth tables

The meaning of an operator or expression can be described by listing all the possible values of the variables in that expression, together with the value of the expression, in a *truth table*. The truth tables for the three basic operators are given below.

A	NOT A ($\overline{A}$)
0	1
1	0

A	B	A AND B ($A.B$)
0	0	0
0	1	0
1	0	0
1	1	1

A	B	A OR B ($A + B$)
0	0	0
0	1	1
1	0	1
1	1	1

In digital design, three further operators are commonly used, NAND (Not AND), NOR (Not OR) and XOR (eXclusive OR).

A	B	A NAND B $(\overline{A.B})$
0	0	1
0	1	1
1	0	1
1	1	0

A	B	A NOR B $(\overline{A + B})$
0	0	1
0	1	0
1	0	0
1	1	0

A	B	A XOR B $(A \oplus B)$
0	0	0
0	1	1
1	0	1
1	1	0

The XNOR $(\overline{A \oplus B})$ operator is also used occasionally. XNOR is the same as EQUIVALENCE.

2.1.4   Rules of Boolean algebra

There are a number of basic rules of Boolean algebra that follow from the precedence of the operators.

1. Commutivity

$A + B = B + A$

$A . B = B . A$

2. Associativity

$A + (B + C) = (A + B) + C$

$A . (B . C) = (A . B) . C$

3. Distributivity

$A . (B + C) = A . B + A . C$

In addition, some basic relationships can be observed from the truth tables above:

$$\overline{\overline{A}} = A$$

$A . 1 = A \qquad A + 0 = A$

$A . 0 = 0 \qquad A + 1 = 1$

$A . A = A \qquad A + A = A$

$A . \overline{A} = 0 \qquad A + \overline{A} = 1$

The right-hand column can be derived from the left-hand column by applying the *principle of duality*. The principle of duality states that if each AND is changed to an OR, each OR to an AND, each 1 to 0 and each 0 to 1, the value of the expression remains the same.

2.1.5   De Morgan's law

There is a very important relationship that can be used to rewrite Boolean expressions in terms of NAND or NOR operations: De Morgan's Law. This is expressed as

$$(\overline{A.B}) = \overline{A} + \overline{B} \quad \text{or} \quad (\overline{A + B}) = \overline{A}.\overline{B}$$

2.1.6   Shannon's expansion theorem

Shannon's expansion theorem can be used to manipulate Boolean expansions.

$$F(A, B, C, D, \dots) = A.F(1, B, C, D, \dots) + \overline{A}.F(0, B, C, D, \dots)$$
$$= (A + F(0, B, C, D, \dots)).(\overline{A} + F(1, B, C, D, \dots))$$

$F(1, B, C, D, \dots)$ means that all instances of A in F are replaced by a logic 1.

2.2   Logic gates

The basic symbols for one and two input logic gates are shown in Figure 2.1. Three or more inputs are shown by adding extra inputs (but note that there is no such thing as a three-input XOR gate). The ANSI/IEEE symbols can be used instead of the traditional 'spade'-shaped symbols, but are 'not preferred' according to IEEE Standard 91-1984. As will be seen in the next chapter, IEEE notation is useful for describing complex logic blocks, but simple sketches are often clearer if done with the traditional symbols. A circle shows logic inversion. Note that there are two forms of the NAND and NOR gates. From De Morgan's law, it can be seen that the two forms are equivalent in each case.

In drawing circuit diagrams, it is desirable, for clarity, to choose the form of a logic gate that allows inverting circles to be joined. The circuits of Figure 2.2 are identical in function. If the circuit of Figure 2.2(a) is to be implemented using NAND gates, the diagram of Figure 2.2(b) may be preferable to that of Figure 2.2(c), because the function of the circuit is clearer.

2.3   Combinational logic design

The values of the output variables of combinational logic are dependent only on the input values and are independent of previous input values or states. Sequential logic, on the other hand, has outputs that depend on the previous states of the system. The design of sequential systems is described in later chapters.

The major design objective is usually to minimize the cost of the hardware needed to implement a logic function. That cost can usually be expressed in terms of the number

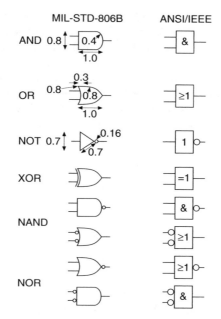

Figure 2.1   Logic symbols.

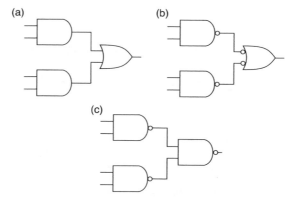

Figure 2.2   Equivalent circuit representations.

of gates, although for technologies such as programmable logic there are other limitations, such as the number of terms that may be implemented. Other design objectives may include testability (discussed in detail in Chapter 11) and reliability.

Before describing the logic design process, some terms have to be defined. In these definitions it is assumed that we are designing a piece of combinational logic with a number of input variables and a single output.

A *minterm* is a Boolean AND function containing exactly one instance of each input variable or its inverse. A *maxterm* is a Boolean OR function with exactly one instance of each variable or its inverse. For a combinational logic circuit with n input variables, there are 2^n possible minterms and 2^n possible maxterms. If the logic function is true at row i of the standard truth table, that minterm exists and is designated by m_i. If the logic function is false at row i of the standard truth table, that maxterm exists and is designated by M_i. For example, the following truth table defines a logic function. The final column shows the minterms and maxterms for the function.

A	B	C	Z	
0	0	0	1	m_0
0	0	1	1	m_1
0	1	0	0	M_2
0	1	1	0	M_3
1	0	0	0	M_4
1	0	1	1	m_5
1	1	0	0	M_6
1	1	1	1	m_7

The logic function may be described by the logical OR of its minterms:

$$Z = m_0 + m_1 + m_5 + m_7$$

A function expressed as a logical OR of distinct minterms is in *sum of products* form:

$$Z = \overline{A}.\overline{B}.\overline{C} + \overline{A}.\overline{B}.C + A.\overline{B}.C + A.B.C$$

Each variable is inverted if there is a corresponding 0 in the truth table and not inverted if there is a 1.

Similarly, the logic function may be described by the logical AND of its maxterms:

$$Z = M_2.M_3.M_4.M_6$$

A function expressed as a logical AND of distinct maxterms is in *product of sums* form:

$$Z = (A + \overline{B} + C)(A + \overline{B} + \overline{C})(\overline{A} + B + C)(\overline{A} + \overline{B} + C)$$

Each variable is inverted if there is a corresponding 1 in the truth table and not inverted if there is a 0.

An *implicant* is a term that covers at least one true value and no false values of a function. For example, the function $Z = A + \overline{A}.\overline{B}$ is shown in the following truth table.

A	B	Z
0	0	1
0	1	0
1	0	1
1	1	1

The implicants of this function are $A.B$, A, $\overline{B}$, $\overline{A}.\overline{B}$ and $A.\overline{B}$. The non-implicants are $\overline{A}$, B and $\overline{A}.B$.

A *prime implicant* is an implicant that covers one or more minterms of a function, such that the minterms are not all covered by another single implicant. In the example above, A and $\overline{B}$ are prime implicants. The other implicants are all covered by one of the prime implicants. An *essential prime implicant* is a prime implicant that covers an implicant not covered by any other prime implicant. Thus, A and $\overline{B}$ are essential prime implicants.

2.3.1   Logic minimization

The function of a combinational logic circuit can be described by one or more Boolean expressions. These expressions can be derived from the specification of the system. It is very likely, however, that these expressions are not initially stated in their simplest form. Therefore, if these expressions were directly implemented as logic gates, the amount of hardware required would not be minimal. Therefore, we seek to simplify the Boolean expressions and hence minimize the number of gates needed. Another way of stating this is to say that we are trying to find the set of prime implicants of a function that is necessary to fully describe the function.

It is possible in principle to simplify Boolean expressions by applying the various rules of Boolean algebra described in Section 2.1. It doesn't take long, however, to realize that this approach is slow and error prone. Other techniques have to be employed. The technique described here, *Karnaugh maps*, is a graphical method, although it is effectively limited to problems with six or fewer variables. The *Quine–McCluskey* algorithm is a tabular method which is not limited in the number of variables and which is well suited to tackling problems with more than one output. Quine–McCluskey can be performed by hand, but it is generally less easy than the Karnaugh map method. It is better implemented as a computer program. Logic minimization belongs, however, to the *NP-complete* class of problems. This means that as the number of variables increases, the time to find a solution increases exponentially. Therefore heuristic methods have been developed that find acceptable, but possibly less than optimal, solutions. The *Espresso* program implements heuristic methods that reduce to the Quine–McCluskey algorithm for small problems. Espresso has been used in a number of logic synthesis systems. Therefore the approach adopted here is to use Karnaugh maps for small problems with a single output and up to six inputs. In general, it makes sense to use an EDA program to solve larger problems.

The Karnaugh map (or K-map, for short) method generates a solution in sum-of-products or product-of-sums form. Such a solution can be implemented directly as two-level AND–OR or OR–AND logic (ignoring the cost of generating the inverse values of inputs). AND–OR logic is equivalent to NAND–NAND logic, and OR–AND logic is equivalent to NOR–NOR logic. Sometimes, a cheaper (in terms of the number of gates) method can be found by factorizing the two-level minimized expression to generate more levels of logic – two-level minimization must be performed before any such factorization. Again, we shall assume that if such factorization is to be performed it will be done using an EDA program, such as *SIS*.

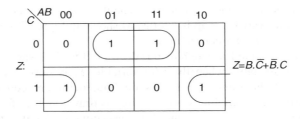

Figure 2.7   Groupings on three-input Karnaugh map.

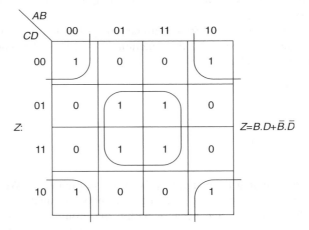

Figure 2.8   Groupings on four-input Karnaugh map.

- Circle 1s and read the sum of products for Z.
- Circle 0s and read the sum of products for $\overline{Z}$.
- Circle 0s and read the product of sums for Z.
- Circle 1s and read the product of sums for $\overline{Z}$.

Diagonal pairs, as shown in Figure 2.9, correspond to XOR functions.

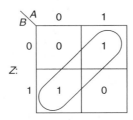

Figure 2.9   Exclusive OR grouping on Karnaugh map.

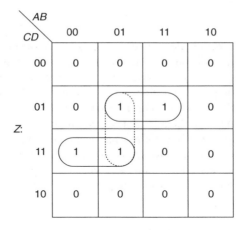

Figure 2.10 Redundant grouping on Karnaugh map.

The Karnaugh map of Figure 2.10 has three prime implicants circled. The function can be read as $Z = B.\overline{C}.D + \overline{A}.C.D + \overline{A}.B.D$. The vertical grouping, shown with a dashed line, covers 1s covered by the other groupings. This grouping is therefore *redundant* and can be omitted. Hence the function can be read as $Z = B.\overline{C}.D + \overline{A}.C.D$.

Assuming that all the prime implicants have been correctly identified, the minimal form of the function can be read by selecting all the essential prime implicants (i.e. those circles that circle 1s – or 0s – not circled by any other group), together with sufficient other prime implicants needed to cover all the 1s (or 0s). Redundant groupings can be ignored, but under some circumstances it may be desirable to include them.

Incompletely specified functions have 'don't cares' in the truth tables. These don't cares correspond to input combinations that will not (or should not) occur. For example, consider the truth table of Figure 2.11. The don't care entries can be included or excluded from groups as convenient, in order to get the largest possible groupings, and hence the smallest number of implicants. In the example, we could treat the don't care as a 0 and read $Z = \overline{A}.\overline{B} + A.B$, or treat the don't care as a 1 and read $Z = \overline{A} + B$.

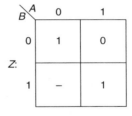

Figure 2.11 Don't care on Karnaugh map.

2.4 Timing

The previous section dealt with minimizing Boolean expressions. The minimized Boolean expressions can then be directly implemented as networks of gates or on programmable logic. All gates have a finite delay between a change at an input and the change at an output. If gates are used, therefore, different paths may exist in the network, with different delays. This may cause problems.

To understand the difficulties, it is helpful to draw a *timing diagram*. This is a diagram of the input and output waveforms as a function of time. For example, Figure 2.12 shows the timing diagram for an inverter. Note the stylized (finite) rise and fall times. An arrow shows causality, i.e. the fact that the change in the output results from a change in the input.

A more complex circuit would implement the function

$$Z = A.C + B.\overline{C}$$

The value of $\overline{C}$ is generated from C by an inverter. A possible implementation of this function is therefore given in Figure 2.13. In practice, the delay through each gate and

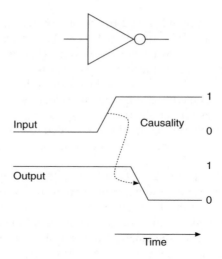

Figure 2.12 Timing diagram for inverter.

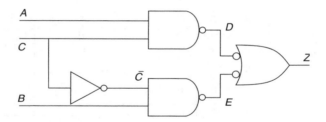

Figure 2.13 Circuit with Static 1 hazard.

through each type of gate would be slightly different. For simplicity, however, let us assume that the delay through each gate is one unit of time. To start with, let $A = 1$ and $B = 1$. The output, Z, should be at 1 irrespective of the value of C. Let us see, by way of the timing diagram in Figure 2.14, what happens when C changes from 1 to 0. One unit of time after C changes, $\overline{C}$ and D change to 1. In turn, these changes cause E and Z to change to 0 another unit of time later. Finally, the change in E causes Z to change back to 1 a further unit of time later. This change in Z from 1 to 0 and back to 1 is known as a *hazard*. A hazard occurs as a result of delays in a circuit.

Figure 2.15 shows the different types of hazard that can occur. The hazard in the circuit of Figure 2.13 is a Static 1 hazard. Static 1 hazards can occur in AND–OR or

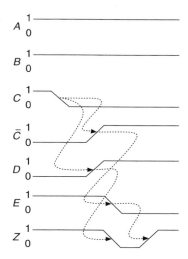

Figure 2.14   Timing diagram for circuit of Figure 2.13.

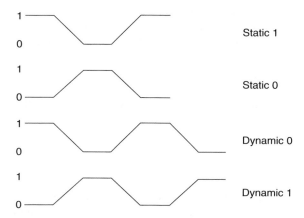

Figure 2.15   Types of hazard.

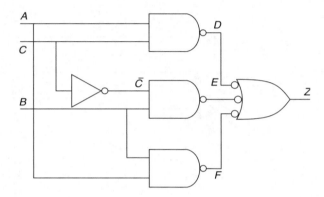

Figure 2.16   Redundant term on Karnaugh map.

Figure 2.17   Hazard-free circuit.

NAND–NAND logic. Static 0 hazards can occur in OR–AND or NOR–NOR logic. Dynamic hazards do not occur in two-level circuits. They require three or more unequal signal paths. Dynamic hazards are often caused by poor factorization in multi-level minimization.

Static hazards, on the other hand, can be avoided by designing with redundant logic. For example, the Karnaugh map of the circuit function of Figure 2.13 is shown in Figure 2.16. The redundant prime implicant is shown as a dotted circle. The redundant gate corresponding to this prime implicant can be introduced to eliminate the hazard. The circuit function is therefore

$$Z = A.C + B.\overline{C} + A.B$$

The circuit is shown in Figure 2.17. Now, F is independent of C. If $A = B = 1$, $F = 0$. F stays at 0 while C changes, therefore Z stays at 1.

2.5 Number codes

Digital signals are either control signals of some kind or information. In general, information takes the form of numbers or characters. These numbers and characters have to be coded in a form suitable for storage and manipulation by digital hardware. Thus one integer or one character may be represented by a set of bits. From the point of view of a computer or other digital system, no one system of coding is better than another. There do, however, need to be standards, so that different systems can communicate. The standards that have emerged are generally also designed such that a human being can interpret the data if necessary.

2.5.1   Integers

The simplest form of coding is that of positive integers. For example, a set of three bits would allow us to represent the decimal integers 0 to 7. In base 2 arithmetic, 000_2 represents 0_{10}, 011_2 represents 3_{10} and 111_2 represents 7_{10}. As with decimal notation, the most significant bit is on the left.

For the benefit of human beings, strings of bits may be grouped into sets of three or four and written using *octal* (base 8) or *hexadecimal* (base 16) notation. For example, 66_8 is equal to $110\ 110_2$ or 54_{10}. For hexadecimal notation, the letters A to F represent the decimal numbers 10 to 15. For example, EDA_{16} is $1110\ 1101\ 1010_2$ or 7332_8 or 3802_{10}.

The simple translation of a decimal number into bits is sufficient for zero and positive integers. Negative integers require additional information. The simplest approach is to set aside one bit as a sign bit. Therefore, $0\ 110_2$ might represent $+6_{10}$, while $1\ 110_2$ would represent -6_{10}. While this makes translation between binary and decimal numbers simple, the arithmetic operations of addition and subtraction require that the sign bits be checked before an operation can be performed on two numbers. It is common, therefore, to use a different notation for signed integers: *two's complement*. The principle of two's complement notation is that the code for $-b$, where b is a binary number represented using n bits, is the code given by $2^n - b$. For example, -6_{10} is represented by $10000_2 - 0110_2$, which is 1010_2. The same result is obtained by inverting all the bits and adding 1: -6_{10} is $1001_2 + 1 = 1010_2$.

The advantage of two's complement notation is that addition and subtraction may be performed using exactly the same hardware as for unsigned arithmetic; no sign checking is needed. The major disadvantage is that multiplication and division become much more complicated. Booth's algorithm, described in Section 6.7, is a technique for multiplying two's complement numbers.

2.5.2   Fixed-point numbers

For many applications, non-integer data need to be stored and manipulated. The binary representation of a *fixed-point* number is exactly the same as for an integer number, except that there is an implicit 'decimal' point. For example, 6.25 is equal to $2^2 + 2^1 + 2^{-2}$ or 110.01_2. Instead of representing the point, the number 11001_2 (25_{10}) is stored, with the implicit knowledge that it and the results of any operations involving it have to be divided by 2^2 to obtain the true value. Notice that all operations, including those for two's complement representations, are the same as for integer numbers.

2.5.3   Floating-point numbers

The number of bits that have been allocated to represent fractions limits the range of fixed-point numbers. *Floating-point* numbers allow a much wider range of accuracy. In general, floating-point operations are only performed using specialized hardware, because they are computationally very expensive. A typical *single-precision* floating-point number has 32 bits, of which one is the sign bit (s), eight are the exponent (e) in two's complement form, and the remaining 23 are the mantissa (m), such that a decimal number is represented as

$$(-1)^s \times 1.m \times 2^e$$

The IEEE standard 754-1985 defines formats for 32, 64 and 128 bit floating-point numbers, with special patterns for $\pm\infty$ and the results of invalid operations, such as $\sqrt{-1}$.

2.5.4   Alphanumeric characters

Characters are commonly represented by seven or eight bits. The ASCII code is widely used. Seven bits allow the basic Latin alphabet in upper and lower cases, together with various punctuation symbols and control codes, to be represented. For example, the letter A is represented by 1000001. For accented characters eight-bit codes are commonly used. Manipulation of text is normally performed using general-purpose computers rather than specialized digital hardware.

2.5.5   Gray codes

In the normal binary counting sequence, the transition from 0111 (7_{10}) to 1000 (8_{10}) causes three bits to change. In some circumstances, it may be undesirable that several bits should change at once, because the bit changes may not occur at exactly the same time. The intermediate values might generate spurious warnings. A Gray code is one in which only one bit changes at a time. For example a three-bit Gray code would count through the following sequence (other Gray codes can also be derived):

        000
        001
        011
        010
        110
        111
        101
        100

Note that the sequence of bits on a K-map is a Gray code. Another application of Gray codes is as a position encoder on a rotating shaft, as shown in Figure 2.18.

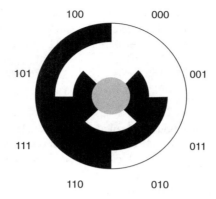

Figure 2.18 Gray code as shaft encoder.

2.5.6 Parity bits

When data are transmitted, either by wire or by using radio communications, there is always the possibility that noise may cause a bit to be misinterpreted. At the very least it is desirable to know that an error has occurred, and it may be desirable to transmit sufficient information to allow any error to be corrected.

The simplest form of error detection is to use a parity bit with each word of data. For each eight bits of data, a ninth bit is sent that is 0 if there are an even number of ones in the data word (even parity) or 1 otherwise. Alternatively odd parity can be used, in which case the parity bit is inverted. This is sufficient if the chances of an error occurring are low. We cannot tell which bit is in error, but knowing that an error has occurred means that the data can be transmitted again. Unfortunately, if two errors occur, the parity bit might appear to be correct. A single error can be corrected by using a two-dimensional parity scheme in which every ninth word is itself a set of parity bits, as shown below. If a single error occurs, both the row parity and column parity will be incorrect, allowing the erroneous bit to be identified and corrected. Certain multiple errors are also detectable and correctable.

	Bit 7	Bit 6	Bit 5	Bit 4	Bit 3	Bit 2	Bit 1	Bit 0	Parity
Word 0	0	1	0	1	0	1	1	0	0
Word 1	0	1	0	0	1	0	0	0	0
Word 2	0	1	0	1	0	0	1	1	0
Word 3	0	1	0	0	1	0	0	1	1
Word 4	0	1	0	0	0	0	1	1	1
Word 5	0	1	0	0	1	0	0	0	0
Word 6	0	1	0	0	0	1	0	0	0
Word 7	0	1	0	0	1	1	0	0	1
Parity	0	0	0	0	0	1	1	1	1

By using a greater number of parity bits, each derived from part of the word, multiple errors can be detected and corrected. The simplest forms of such codes were derived by Hamming in 1948. Better codes were derived by Reed and Solomon in 1960.

Summary

Digital design is based on Boolean algebra. The rules of Boolean algebra allow logical expressions to be simplified. The basic logical operators can be implemented as digital building blocks – gates. Graphical methods, Karnaugh maps, are a suitable tool for finding the minimal forms of Boolean expressions with fewer than six variables. Larger problems can be tackled with computer-based methods. Gates have delays, which means that non-minimal forms of Boolean expressions may be needed to prevent timing problems, known as hazards. Data can be represented using sets of bits. Different types of data can be encoded to allow manipulation. Error-detecting codes are used when data is transmitted over radio or other networks.

Further reading

The principles of Boolean algebra and Boolean minimization are covered in many books on digital design. Recommended are those by Wakerly and by Hill and Peterson. De Micheli describes the Espresso algorithm, which sits at the heart of many logic optimization software packages. Espresso may be downloaded from http://www-cad.eecs.berkeley.edu/

Error detection and correction codes are widely used in communications systems. Descriptions of these codes can be found in, for example, Hamming.

Exercises

2.1 Derive Boolean expressions for the circuits of Figure 2.19; use truth tables to discover if they are equivalent.

2.2 Minimize

(a) $Z = m_0 + m_1 + m_2 + m_5 + m_7 + m_8 + m_{10} + m_{14} + m_{15}$

(b) $Z = m_3 + m_4 + m_5 + m_7 + m_9 + m_{13} + m_{14} + m_{15}$

2.3 Describe two ways of representing negative binary numbers. What are the advantages and disadvantages of each method?

2.4 A floating-point decimal number may be represented as:

$(-1)^s \times 1.m \times 2^e$

Explain what the binary numbers s, m and e represent. How many bits would typically be used for s, m and e in a single-precision floating-point number?

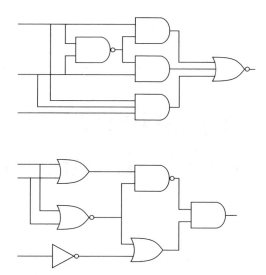

Figure 2.19   Circuits for Exercise 2.1.

Chapter 3

Combinational logic using VHDL gate models

Combinational logic is stateless: changes in inputs are immediately reflected by changes in outputs. In this chapter we will introduce the basic ideas of modelling in VHDL by looking at combinational logic described in terms of gates.

3.1 Entities and architectures

Even the most basic VHDL model has two parts: an **entity** and an **architecture**. For example, a two-input AND gate may be described by:

```
entity And2 is
  port (x, y : in BIT; z: out BIT);
end entity And2;

architecture ex1 of And2 is
begin
  z <= x and y;
end architecture ex1;
```

38

The **entity** part describes a black box. We can see the inputs and outputs of the black box, together with their types, but we know nothing of the internals of the circuit. The **architecture** describes the function and/or structure of the circuit. In this example, the functionality of the circuit is described in terms of Boolean operations. The reason for having this split is that it is possible to have more than one architecture for each entity, perhaps describing alternative implementations or different levels of description. For instance, we can describe an AND gate in terms of a Boolean operator, as shown, but we could also write a truth table. In either case, the entity, i.e. the 'black box', is the same, but there would be two architectures, one for each model.

The words shown in **bold** are reserved words. The entity description starts with the reserved word **entity**, the name of the entity and the reserved word **is**. The entity finishes with the reserved words **end entity** and the entity name. The entity name at the end is optional, but if included must be the same as that used in the first line of the entity declaration. It is, however, strongly recommended that you include all names after an **end** for clarity. Note that the entity declaration concludes with a semicolon ('**;**'). Semicolons are used to mark the end of statements.

Words shown in UPPERCASE are built-in types or other identifiers. In later chapters, we will introduce standard libraries. Identifiers from those libraries will not be shown in uppercase. You should treat any identifiers shown in uppercase as reserved words. In other words if you redefine them, you may have problems.

The original VHDL standard was defined in 1987. The 1993 standard introduced some new features and made the syntax of many VHDL constructs much more consistent (and longer). We will use the longer forms of the 1993 standard throughout this book, in order to make the examples more readable. In the 1987 VHDL standard, **end** rather than **end entity** was used. This also applied to various other structures. To ensure backward compatibility, it is possible to miss out the reserved word **entity** between **end** and the name. A further revision was agreed in 2002. The differences between the 1993 and 2002 standards are mostly minor. Appendix A has a summary of the differences between the 1987, 1993 and 2002 standards.

The entity contains one declaration. The reserved word **port** is used to specify connections between the entity and the outside world. Here, two signals of type BIT (x and y) are defined by the reserved word **in** to be inputs to the model and one signal of type BIT is an output. The reserved word **in** is optional (i.e. unless otherwise stated, all signals at ports are inputs). Bidirectional signals are indicated by the reserved word **inout**. BIT is a predefined type with two values: '0' and '1'.

The architecture declaration includes its own name and that of the entity with which it is associated. The entity name must refer to a previously defined entity. In this example, the model has only one statement between the reserved word **begin** and the **end architecture** line. Signal z is *assigned* the value of x **and** y. **And** is a built-in operator, which takes operands of type BIT and returns a result of type BIT.

VHDL has the following operators defined for type BIT: **not**, **and**, **or**, **nand**, **nor**, **xor** and **xnor**. The **not** operator has the highest precedence (i.e. it is evaluated first). The remaining operators all have the same precedence, and are evaluated in the order in which they are written, left to right. Note that Boolean algebra normally defines the AND operator to have a higher precedence than the OR operator: extra care must be taken when writing VHDL to ensure that expressions are interpreted correctly.

3.2 Identifiers, spaces and comments

VHDL is not case-sensitive (unlike C). Thus, '**architecture**', '**ARCHITECTURE**' and '**aRcHiTeCtUrE**' are all equivalent and acceptable *to a compiler*. Similarly, identifiers (such as 'And2') may be mixed-case. It is strongly recommended, however, that the usual software engineering rules about identifiers should be applied:

- Meaningful, non-cryptic names should be used, based on English words.
- Use mixed-case with consistent use of case.
- Don't use excessively long identifiers (use 15 characters or fewer).
- Don't use identifiers that may be confused (e.g. two identifiers that differ by an underscore).
- Don't redefine predefined identifiers, such as BIT or TIME.
- Identifiers may consist of letters, numbers and underscores ('_'), but the first character must be a letter, and two underscores in succession are not allowed. Extended identifiers may consist of any character, provided that the entire identifier is enclosed in backslashes ('\'), e.g. \0%$#___&\. The strings in extended identifiers are case-sensitive. Use extended identifiers with extreme caution.

Whitespace (spaces, carriage returns) should be used to make models more readable. There is no difference between one whitespace character and many.

Comments may be included in a VHDL description by putting two hyphens on a line ('--'). All text between the hyphens and the end of the line is ignored. This is similar to the C++ style of comment, which uses two slashes ('//'). There is no C-style block comment ('/* . . . */') in VHDL. It is strongly recommended that comments should be included to aid in the understanding of VHDL code. Each VHDL file should include a header, which typically contains:

- the name(s) of the design units in the file;
- file name;
- a description of the code;
- limitations and known errors;
- any operating system and tool dependencies;
- the author(s), including a full address;
- a revision list, including dates.

For example:

```
-----------------------------------------------------------
-- Design unit : And2(Example) (Entity and Architecture)
--              :
-- File name   : and2.vhd
--              :
-- Description : Dataflow model of basic 2 input and
--              : gate. Inputs of type BIT.
```

```
-- Limitations : None
--             :
-- System      : VHDL'93
--             :
-- Author      : Mark Zwolinski
--             : Department of Electronics and Computer
--             : Science
--             : University of Southampton
--             : Southampton SO17 1BJ, UK
--             : mz@ecs.soton.ac.uk
--
-- Revision    : Version 1.0 04/02/99
---------------------------------------------------------------
```

3.3 Netlists

We have seen in Section 3.1 how to describe a two-input AND gate. Combinational logic is seldom that simple! Suppose we wish to build a circuit to implement the following truth table:

A	B	C	Z
0	0	0	0
0	0	1	0
0	1	0	1
0	1	1	1
1	0	0	0
1	0	1	1
1	1	0	0
1	1	1	1

From a K-map, we can deduce that a minimal form of this function is

$$Z = \overline{A}.B + A.C$$

Therefore a VHDL implementation of this function might be:

```
entity comb_function is
  port (a, b, c : in BIT; z: out BIT);
end entity comb_function;
```

```
architecture expression of comb_function is
begin
   z <= (not a and b) or (a and c);
end architecture expression;
```

Note the use of parentheses to ensure the correct order of evaluation – **and** and **or** have the same precedence, as described earlier.

While this model is perfectly acceptable as a description of a particular logical function, it does not correspond to any normally available piece of circuitry. It could be implemented as a PLA function, but if we wished to build this function on an integrated circuit using only standard logic gates we would need to rewrite the model in terms of those gates. This might be done using some EDA program, but let us see how we can do this manually.

Let us write models of an `Or2` gate and a `Not1` gate in addition to the `And2` gate shown earlier.

```
entity Or2 is
  port (x, y : in BIT; z: out BIT);
end entity Or2;

architecture ex1 of Or2 is
begin
  z <= x or y;
end architecture ex1;

entity Not1 is
  port (x : in BIT; z: out BIT);
end entity Not1;

architecture ex1 of Not1 is
begin
  z <= not x;
end architecture ex1;
```

There is no conflict in calling each of the architectures 'ex1' as each applies to a different entity.

A VHDL description of the function above using these gates might be written as:

```
architecture netlist of comb_function is
  signal p, q, r : BIT;
begin
  g1: entity WORK.Not1(ex1) port map (a, p);
  g2: entity WORK.And2(ex1) port map (p, b, q);
  g3: entity WORK.And2(ex1) port map (a, c, r);
  g4: entity WORK.Or2(ex1) port map (q, r, z);
end architecture netlist;
```

Several new pieces of VHDL are introduced here. Within the architecture, we need to create instances of each gate. We also need to identify the signals that connect the gates to the rest of the world.

To identify the signals connected to each gate within the model, we have a **signal** declaration, which simply lists the names of the signals and their types.

To create an instance of a gate, we include a reference to it between the **begin** and **end** in the architecture. Generally, in VHDL objects must be declared before they are used. Here, however, we do not declare each gate explicitly but we do have to define exactly where to find each model and which architecture to use.

'WORK' refers to the current working library. As each entity and architecture is compiled, it is stored in a directory. We don't want to know exactly which operating system directory or file is to be used; thus WORK is an operating system-independent way of specifying that the compiler should look in the current working directory. This is good software engineering practice; we should avoid writing VHDL that is specific to a particular EDA tool or system. '(ex1)' refers, in each case, to the architecture of each gate model. This style of hierarchy description is called *direct instantiation*.

The instance has a name (g1), the type of the gate (Not1) and a clause showing how external signal names are mapped to internal signal names. The reserved words **port map** are used to show these mappings. Note that, as shown, the order of signals is important. We can assign signals in a different order:

```
g2: entity WORK.And2(ex1) port map (z => q, x => p, y =>b);
```

The direct instantiation style is usually preferred for simple netlists. An alternative style has the components declared before they are used:

```
architecture netlist2 of comb_function is
  component And2 is
    port (x, y : in BIT; z: out BIT);
  end component And2;
  component Or2 is
    port (x, y : in BIT; z: out BIT);
  end component Or2;
  component Not1 is
    port (x : in BIT; z: out BIT);
  end component Not1;
  signal p, q, r : BIT;
begin
  g1: Not1 port map (a, p);
  g2: And2 port map (p, b, q);
  g3: And2 port map (a, c, r);
  g4: Or2 port map (q, r, z);
end architecture netlist2;
```

The **component** declaration tells the compiler what each gate looks like. The component declaration is identical to an entity declaration, with the reserved word **entity** replaced by **component**. Here, the component declaration has the same name as the entity declaration and the ports also have the same names. We also assume here that there is only one architecture for each gate, or more accurately, that if more than one architecture exists, we will use the last one compiled. This is known as the *default configuration*. Later, we will see how different architectures can be explicitly mapped to particular instances, and how we can associate entities and components with different names.

3.4   Signal assignments

The And2 gate has a *signal assignment* of the form:

```
z <= x and y;
```

In other words, a signal z takes the value of the logical AND of two signals x and y. In its simplest form a signal assignment passes the value of one signal directly to another:

```
z <= x;
```

This can be enhanced to provide more complex logical operations:

```
z <= not ((x and y) or (a and b));
```

Thus it is possible to describe a wide variety of gates and other combinational logic, simply using signal assignments.

In reality, of course, the output of a piece of real hardware does not change instantaneously when an input changes. Inevitably, there is a delay. If we were to model the circuit in terms of circuit components, such as transistors, the cause of delays would be obvious – transistors have capacitances; interconnects can be thought of as transmission lines, etc. In digital design, we think of circuits in terms of gates and other similar building blocks. Therefore we choose to lump all the delay elements together and describe the propagation of a signal in terms of a single delay. For example, we can delay a signal assignment by, say, 4 ns, using the following construct in VHDL:

```
z <= x after 4 NS;
```

This is an *inertial* delay. In other words, the signal is delayed by 4 ns, and in addition, any pulse that is less than 4 ns wide is suppressed, as shown in Figure 3.1.

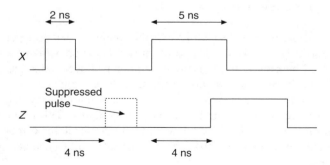

Figure 3.1   Inertial delay of 4 ns.

A pure or transport delay is modelled with:

```
z <= transport x after 4 NS;
```

(Note the space between 4 and NS.) Any pulse is now transmitted. We can include the keyword **inertial** if we want to be very specific about the delay model.

The delay construct can be applied to logical expressions. For example:

```
z <= x and y after 5 NS;
```

```
z <= transport not ((x and y) or (a and b)) after 8 NS;
```

Multiple changes are allowed in response to a single signal change:

```
z <= x after 4 NS, not x after 8 NS;
```

would set the value of z in response to a change in x and then set a new value 4 ns after the first. Although this type of multiple change is not particularly useful for modelling combinational logic, the construct has other uses, as will be seen in Section 3.7.

At this point it can be seen that we can describe gates and other combinational logic in terms of their logical functions and in terms of their behaviour in time. We will see later the restrictions that are placed upon VHDL constructs that are intended for synthesis, but we can already draw an important conclusion about synthesizable models. The expression $Z = \overline{A}.B + A.C$ describes a logical function. It is possible to implement this in a number of ways, using NAND gates, programmable logic, etc. Whatever implementation we choose, the logical function is achieved. On the other hand, if we define the delay through the gate to be 5 ns, we are stating the behaviour of a particular implementation.

3.5 Generics

The statement:

```
z <= x and y after 5 ns;
```

defines the exact delay for an AND gate. Different technologies, and indeed different instances, will have different delays. We could declare a number of alternative architectures for an AND gate, each with a different delay. It would be better to write the statement as:

```
z <= x and y after delay;
```

and to define delay as a parameter to the VHDL model. This is achieved using a **generic**:

```
entity And2 is
  generic (delay : DELAY_LENGTH);
  port (x, y : in BIT; z: out BIT);
end entity And2;
```

```
architecture ex2 of And2 is
begin
  z <= x and y after delay;
end architecture ex2;
```

When the gate is used in a netlist, a value is passed to the model using a **generic map**:

```
g2: entity WORK.And2(ex2) generic map (5 NS)
                          port map (p, b, q);
```

A **component** declaration must also include the **generic** declaration. The non-positional form can also be used:

```
g2: entity WORK.And2(ex2) generic map (delay => 5 NS)
                          port map (z => q, x =>p, y =>b);
```

As we will see later, generics are also useful for passing structural information, for instance how many bits there are in an adder.

It can be useful to specify a default value for a generic. This allows the generic map to be omitted if the default value is to be used. For example, the delay through the And2 gate might normally be 5 ns, but occasionally a gate with a different delay might exist. Default values are specified as follows:

```
generic (delay : DELAY_LENGTH := 5 NS);
```

Therefore if the component is instantiated as

```
g2: entity WORK.And2(ex2) port map (p, b, q);
```

a value of 5 ns will be passed to the delay. A different value can override the default if it is explicitly stated.

```
g2: entity WORK.And2(ex2) generic map (3 NS)
                          port map (p, b, q);
```

If a component declaration is used, the default value can be specified in the **entity** declaration or the **component** declaration or in both. If different default values are specified in each declaration, the value in the component declaration will be used. If no default value is given and the **generic map** part is omitted, delay would be undefined, so an error will be generated. If the component declaration does not include the generic definition, the default value will automatically be used and the **generic map** part of the **component** instantiation must be omitted.

Finally, the default value of a generic will be used if the reserved word **open** is used as the actual value:

```
g2: entity WORK.And2(ex2) generic map (open)
                          port map (p, b, q);
```

In this case, the entire **generic map** part of the instantiation is redundant, but if there were several generics the reserved word **open** could be used to allow some generics to take default values while others were given specific values.

3.6 Constant and open ports

There may be occasions on which not all the inputs or outputs of a component are needed. Therefore these inputs and outputs would be connected to the supply rails or left unconnected. To illustrate this, let us invent a 'universal' gate. This gate has three inputs and two outputs. The two outputs are the AND and OR functions. Two of the inputs are the normal logical inputs for an AND or OR function. The third input indicates whether the outputs are to be inverted. Thus the gate implements the AND, OR, NAND and NOR functions. A VHDL description of this gate follows.

```
entity universal is
  port (x, y, invert : in BIT; a, o : out BIT);
end entity universal;

architecture univ of universal is
begin
  a <= (y and (x xor invert)) or (invert and not y);
  o <= (not x and (y xor invert)) or (x and not invert);
end architecture univ;
```

It is left as an exercise for the reader to derive the logic equations. Note that the two signal assignments occur concurrently, not sequentially. We will return to this in the next chapter. To use this gate as an AND gate, we would set the invert input to '0' and leave the o output unconnected. This is done using a value instead of a signal and by using the reserved word **open** again:

```
u0 : entity WORK.universal(univ) port map (x, y, '0', a,
                                           open);
```

Outputs can be left unconnected, but inputs may be left **open** only if a default value has been specified in the **entity** declaration or the **component** declaration, as for **generics**:

```
entity universal is
  port (x, y : in BIT;
        invert : in BIT := '0';
        a, o : out BIT);
end entity universal;
```

The following instantiation would now be legal:

```
u0 : entity WORK.universal port map (x, y, open, a, open);
```

3.7   ## Testbenches

If we wish to simulate our circuit to verify that it really does work as expected, we need
to apply some test stimuli. We could, of course, write out some test vectors and apply
them, or, more conveniently, write the test data in VHDL. This type of VHDL model is
often known as a *testbench*. Testbenches have a distinctive style. Below is a testbench
for a two-input AND gate.

```vhdl
entity TestAnd2 is
end entity TestAnd2;

architecture io of TestAnd2 is
   signal a,b,c : BIT;
begin
   g1: entity WORK.And2(ex2) port map (x=>a, y=>b, z=>c);
   a<= '0', '1' after 100 NS;
   b<= '0', '1' after 150 NS;
end architecture io;
```

Because this is a testbench, i.e. a description of the entire world that affects the model
we are testing, there are no inputs or outputs in the entity. This is characteristic of testbenches. The description in the architecture consists of an instance of the circuit we are
testing, together with a set of input stimuli. Signals corresponding to the input and output ports of the circuit are also declared. Inside the body of the architecture one
instance of the circuit is created.

   This is a very simple example of a testbench. It provides sufficient inputs to run a
simulation, but the designer would need to look at the simulation results to check that
the circuit was functioning as intended. VHDL has the richness of a programming language. Therefore a testbench could be written to check simulation results against a file
of expected responses or to compare two versions of the same circuit.

3.8   ## Configurations

Here is another description of a two-input AND gate in VHDL:

```vhdl
architecture ex3 of And2 is
   signal xy : BIT_VECTOR(0 to 1);
begin
   xy <= x&y;
   with xy select
     z <= '1' when "11",
          '0' when others;
end architecture ex3;
```

We will explain the constructs in the next chapter. We now have two different architectures (ex1 from Section 3.1 and ex3) associated with the same **entity**. (ex2 from Section 3.5 is the same as ex1, but with a delay, and hence has a different entity declaration.) In the previous examples, the architecture name was explicitly stated, but it can be omitted. If an entity has only one architecture, there is no ambiguity – both the following have the same meaning:

```
g1: entity WORK.Not1(ex1) port map (a, p);

g1: entity WORK.Not1 port map (a, p);
```

If, however, there were two or more architectures in the same file, then in the second case, by default, we would automatically use the last architecture. Suppose that we wish to have more control over exactly which architecture to use. With direct instantiation, there is no difficulty. With the alternative style, however, there needs to be an explicit statement – the *configuration specification*. For example, the testbench example of the last section could be written as:

```
architecture alternate of TestAnd2 is
  component A2 is
    port (x, y : in BIT; z: out BIT);
  end component A2;
  for all : A2 use entity WORK.And2(ex2);
  signal a,b,c : BIT;
begin
  g1: A2 port map (x=>a, y=>b, z=>c);
end architecture alternate;
```

By using the **for . . . use** construct we can choose which architecture to use. With simple testbenches, the style shown above may be appropriate. For complex models, with several levels of hierarchy, it is often more appropriate to use a **configuration** unit. A configuration declaration for the original testbench shown might be:

```
configuration Tester1 of TestAnd2 is
  for io
    for g1 : And2
      use entity WORK.And2(ex1);
    end for;
  end for;
end configuration Tester1;
```

The complete model therefore consists of the entity and architecture of the And2 gate, the entity and architecture of the testbench and the configuration. There are other ways to write configurations, but this style requires one configuration for the entire design. Note that now we would not include a **for . . . use** statement within the testbench.

It is also possible to use configurations to map **port** and **generic** names. Suppose the testbench were written as:

```
architecture remapped of TestAnd2 is
  component MyAnd2 is
    generic (dly : DELAY_LENGTH);
    port (in1, in2 : in BIT; out1: out BIT);
  end component MyAnd2;
  signal a,b,c : BIT;
begin
  g1: MyAnd2 generic map (6 NS) port map (a, b, c);
end architecture remapped;
```

We would write the configuration as:

```
configuration Tester2 of TestAnd2 is
  for remapped
    for g1 : MyAnd2
      use entity WORK.And2(ex2)
        generic map (delay => dly);
        port map (x => in1, y => in2, z => out1);
    end for;
  end for;
end configuration Tester2;
```

This is a 'board–socket–chip' analogy, where the configuration is used to map between arbitrary internal and external names.

A different style of configuration has one configuration per entity, e.g.:

```
configuration And2Con of And2 is
  for ex1
  end for;
end configuration And2Con;
```

This selects the **architecture** for an **entity**. For the testbench example we then have a **configuration** such as:

```
configuration Tester3 of TestAnd2 is
  for remapped
    for g1 : MyAnd2
      use configuration WORK.And2Con;
    end for;
  end for;
end configuration Tester3;
```

This approach requires a greater number of configuration units, but each unit is simpler. Configurations are important for controlling projects involving a number of designers. For designs done by a single designer using a single FPGA, either configuration statements in each architecture or a single configuration unit is likely to be sufficient.

Summary

A VHDL model has an **entity** part, which is a description of the interface of the model, and one or more **architecture** parts, which describe the functionality of the model. VHDL models should use meaningful identifiers and include comments. In this respect, writing good VHDL is much like writing good software. Netlists of VHDL models can be constructed by instantiating those models. There are a number of alternative ways to instantiate models. Parameters may be passed to models using **generic**s. The reserved word **open** is used to specify an unconnected port or defaulted generic. VHDL models may be exercised using testbenches, also written in VHDL. **Configuration** statements and units are used to associate architectures with particular instances of models.

Further reading

The definition of VHDL is contained in the Language Reference Manual (LRM). This can be bought from the IEEE. Every college or university library should have a copy! There are a number of VHDL books available, but even some recent editions cover only the 1987 standard.

Exercises

3.1 Why does VHDL have entities and architectures?

3.2 What is a configuration used for?

3.3 Write a model of a three-input NAND gate with an inertial delay of 5 ns.

3.4 Write a model of a three-input NAND gate with a parameterizable transport delay.

3.5 A full adder has the following truth table for its sum (S) and carry (Co) outputs, in terms of its inputs, A, B and carry in (Ci):

A	B	Ci	S	Co
0	0	0	0	0
0	0	1	1	0
0	1	0	1	0
0	1	1	0	1
1	0	0	1	0
1	0	1	0	1
1	1	0	0	1
1	1	1	1	1

Derive expressions for S and Co using only AND and OR operators. Hence write a VHDL description of a full adder as a netlist of AND and OR gates and inverters. Do not include any gate delays in your models.

3.6 Write a VHDL testbench to test all combinations of inputs to the full adder of Exercise 3.5. Verify the correctness of your full adder and of the testbench using a VHDL simulator.

3.7 Modify the gate models of Exercise 3.5 such that each gate has a delay of 1 ns. What is the maximum delay through your full adder? Verify this delay by simulation.

Chapter 4

Combinational building blocks

While it is possible to design all combinational (and indeed sequential) circuits in terms of logic gates, in practice this would be extremely tedious. It is far more efficient, in terms of both the designer's time and the use of programmable logic resources, to use higher level building blocks. If we were to build systems using TTL or CMOS integrated circuits on a printed circuit board, we would look in a catalogue and choose devices to implement standard circuit functions. If we use VHDL and programmable logic we are not constrained to using just those devices in the catalogue, but we still think in terms of the same kinds of circuit functions. In this chapter we will look at a number of combinational circuit functions. As we do so, various features of VHDL will be introduced. In addition, the IEEE dependency notation will also be introduced, allowing us to describe circuits using both graphical and textual representations.

4.1 Three-state buffers

4.1.1 Multi-valued logic

In addition to the normal Boolean logic functions, it is possible to design digital hardware using switches to disconnect a signal from a wire. For instance, we can connect

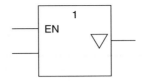

Figure 4.1   Three-state buffer.

the outputs of several gates together, through switches, such that only one output is connected to the common wire at a time. This same functionality could be achieved using conventional logic, but would probably require a greater number of transistors. The IEEE symbol for a three-state buffer is shown in Figure 4.1. The symbol '1' shows the device is a buffer. 'EN' is the symbol for an output enable and the inverted equilateral triangle indicates a three-state output.

When a switched gate is disconnected, it is usual to speak of the output of the entire block, gate and switch, as being in a 'high-impedance' state. This state must be included in the algebra used to define the functionality of gates. We have, so far, used the VHDL type BIT to describe signals. Such signals can take the values of '0' and '1'. If we are going to use the high-impedance state, BIT is no longer adequate to represent logic signal values. We can define a new type to represent logic signals in VHDL:

type tri **is** ('0', '1', 'Z');

Hence we can define signals and ports to be of this type, instead of being of type BIT:

signal a, b, c : tri;

It would obviously now be desirable to be able to use signals of type tri in *exactly* the same way as signals of type BIT. In other words, we want to be able to write VHDL statements of the form:

a <= '0' **and** '1';

b <= a **or** c **after** 5 NS;

Thus we need an **and** operator described by the following truth table. (This assumes that a high-impedance input – a floating input – tends to be pulled to the same value as the other input of an AND gate. This is a modelling decision that may or may not be realistic.) The first row and first column represent the two inputs to the function. The nine elements in the body of the table are the outputs.

AND	0	1	Z
0	0	0	0
1	0	1	1
Z	0	1	1

In VHDL, functions and operators can be *overloaded*. For example, the **and** operator normally takes two operands of type `bit` and returns the Boolean AND of the two operands. We can write a new AND operator to take two operands of type `tri` and return the values shown in the truth table. The syntax of this function will become clear later.

```
function "and" (Left, Right: tri) return tri is
  type tri_array is array (tri, tri) of tri;
  constant and_table : tri_array := (('0', '0', '0'),
                                     ('0', '1', '1'),
                                     ('0', '1', '1'));
begin
  return and_table(Left, Right);
end function "and";
```

The VHDL compiler can work out which is the correct version of the operator to use by the types of the operands. If we tried to AND together a signal of type `BIT` and a signal of type `tri`, the compilation would fail because such an operator has not been defined. We could equally write an **and** operator that implements what is normally considered to be an **or** operator. This can easily render a piece of VHDL code incomprehensible. *Extreme care should be taken with overloading of operators.*

4.1.2 Standard logic package

Having defined a new type with values '0', '1' and 'Z', we would have to write VHDL functions for the various logical operations. Moreover, we might wonder whether three states are sufficient for everything we might wish to model. IEEE standard 1164 defines an enumerated type with nine values:

'U' Uninitialized
'X' Forcing (i.e. strong) unknown
'0' Forcing 0
'1' Forcing 1
'Z' High impedance
'W' Weak unknown
'L' Weak 0
'H' Weak 1
'−' Don't care

The standard logic type is defined by:

```
type std_ulogic is ('U', 'X', '0', '1', 'Z', 'W', 'L',
                    'H', '-');
```

The **and** function for `std_ulogic` is given by the following truth table. As before, the two inputs are given by the first row and column.

	U	X	0	1	Z	W	L	H	–
U	U	U	0	U	U	U	0	U	U
X	U	X	0	X	X	X	0	X	X
0	0	0	0	0	0	0	0	0	0
1	U	X	0	1	X	X	0	1	X
Z	U	X	0	X	X	X	0	X	X
W	U	X	0	X	X	X	0	X	X
L	0	0	0	0	0	0	0	0	0
H	U	X	0	1	X	X	0	1	X
–	U	X	0	X	X	X	0	X	X

If we write a model using signals of type BIT or std_ulogic, we must ensure that two models do not attempt to put a value onto the same signal. In VHDL terms, a signal may have one or more *sources*. A source may be an **out**, **inout** or **buffer** port of an instantiated component or a *driver*. In simple terms, a driver is the right-hand side of a signal assignment. The one occasion when we do try to connect two or more outputs together is when we use three-state buffers. We still have to be careful that no more than one output generates a logic 1 or 0 and the rest of the outputs are in the high-impedance state, but we want the simulator to tell us if there is a design mistake. This cannot be done with std_ulogic – a VHDL simulator does not treat 'Z' as a special case.

The IEEE 1164 standard defines std_logic, which allows more than one output to be connected to the same signal. Std_logic is defined as a **subtype** of std_ulogic, for which a resolution function is declared. The resolved function defines the state of a signal if, for example, a 'Z' and a '1' are driven onto the same signal. Because VHDL is strongly typed, operations involving two or more types must be explicitly defined. A **subtype** may, however, be used in place of the type from which it is derived, without causing an error.

subtype std_logic **is** resolved std_ulogic;

The resolution function is defined by the following truth table.

	U	X	0	1	Z	W	L	H	–
U	U	U	U	U	U	U	U	U	U
X	U	X	X	X	X	X	X	X	X
0	U	X	0	X	0	0	0	0	X
1	U	X	X	1	1	1	1	1	X
Z	U	X	0	1	Z	W	L	H	X
W	U	X	0	1	W	W	W	W	X
L	U	X	0	1	L	W	L	W	X
H	U	X	0	1	H	W	W	H	X
–	U	X	X	X	X	X	X	X	X

Thus a '1' and a '0' driving the same signal would cause that signal to take the value 'X'.

Ideally, we should use `std_ulogic` for all signals unless we intend that any contention should be resolved. If we were to do this, the simulator would immediately tell us (by halting) if we were erroneously trying to force two conflicting values onto the same piece of wire.[1] In practice, however, some synthesis tools have difficulties with `std_ulogic`. The use of `std_logic` now seems to be the accepted industry standard, so in the rest of this book we will use `std_logic` as the types of all Boolean signals. Contention can be recognized by the unexpected appearance of 'X' values in a simulation.

The various standard logic types and the functions needed to use them are gathered together in a **package**. Packages are described in more detail later. It is sufficient to know that a package is a separately compiled set of functions and types. This particular package is kept separately from the working library in a **library** called IEEE. This may translate to a directory somewhere on the system. Therefore, every VHDL model that uses the standard logic package must be prefixed with the lines:

```
library IEEE;
use IEEE.std_logic_1164.all;
```

In general these lines should appear before each entity declaration and will apply to any architectures declared for that entity. If more than one entity declaration appears in a file (for instance, of a model and of its testbench), the **library** and **use** statements must appear before each entity. In other words, VHDL scope rules apply to design units and not to the files in which those design units are declared.

4.1.3   When . . . else statement

The behaviour of a three-state buffer can be described verbally as 'when the enable signal is asserted connect the output to the input, else let the output float'. This statement cannot be implemented using standard logic gates. VHDL has a number of programming constructs to perform this task. One is as follows.

```
library IEEE;
use IEEE.std_logic_1164.all;

entity three_state is
  port (a, enable : in std_logic;
        z : out std_logic);
end entity three_state;

architecture when_else of three_state is
begin
  z <= a when enable = '1' else 'Z';
end architecture when_else;
```

---

[1] In fact, this is exactly what was done in the first edition of this book.

If we wish to model the delay through the buffer, the **when** statement is changed as follows:

```
architecture after_when_else of three_state is
begin
    z <= a after 4 NS when enable = '1' else 'Z';
end architecture after_when_else;
```

4.2   Decoders

4.2.1   2 to 4 decoder

A decoder converts data that has previously been encoded into some other form. For example, n bits can represent 2^n distinct values. The truth table for a 2 to 4 decoder is given below.

Inputs		Outputs			
A1	A0	Z3	Z2	Z1	Z0
0	0	0	0	0	1
0	1	0	0	1	0
1	0	0	1	0	0
1	1	1	0	0	0

The IEEE symbol for a 2 to 4 decoder is shown in Figure 4.2. BIN/1-OF-4 indicates a binary decoder in which one of four outputs will be asserted. The numbers give the 'weight' of each input or output.

We could choose to treat each of the inputs and outputs separately, but as they are obviously related, it makes sense to treat the input and output as two vectors of size 2 and 4 respectively. Vectors can be described using an array of Boolean signals:

```
type std_logic_vector is array (NATURAL range <>) of
    std_logic;
```

NATURAL is a predefined subtype with integer values from 0 to the maximum integer value. **range** <> means an undefined range. Thus std_logic_vector can be used to represent a logic vector of any size, subject to the constraints of the compiler.

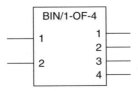

Figure 4.2   2 to 4 decoder.

The 2 to 4 decoder can be modelled using a **when . . . else** statement:

```vhdl
library IEEE;
use IEEE.std_logic_1164.all;

entity decoder is
  port (a : in std_logic_vector(1 downto 0);
        z : out std_logic_vector(3 downto 0));
end entity decoder;

architecture when_else of decoder is
begin
  z <= "0001" when a = "00" else
       "0010" when a = "01" else
       "0100" when a = "10" else
       "1000" when a = "11" else
       "XXXX";
end architecture when_else;
```

The `std_logic_vectors` are declared to have a range of an integer value **downto** zero. It is also possible to declare a range of zero **to** an integer value. In this example, either form would be equally valid. Integers, however, are normally represented such that the most significant bit is on the left. As can be seen from the values of a in the **when** statement, using **downto** rather than **to** allows us to represent integer values in the usual way. The value of a vector is placed within double quotation marks (`"`), unlike that of a single bit.

In this example, there are four **when . . . else** clauses. Each condition (i.e. each value of a) is tested in turn until a condition is found to be true. If none of the conditions is true, i.e. if one bit of a is neither 1 nor 0, the value following the final **else**, corresponding to all bits being unknown, is assigned to z. The final **else** can be omitted:

```vhdl
z <= "0001" when a = "00" else
     "0010" when a = "01" else
     "0100" when a = "10" else
     "1000" when a = "11";
```

If the final **else** is omitted, z continues to take the last value assigned to it. In VHDL, a signal takes a value until a new value is assigned. This may be interpreted as z holding its value in a latch. This is equivalent to writing:

```vhdl
z <= "0001" when a = "00" else
     "0010" when a = "01" else
     "0100" when a = "10" else
     "1000" when a = "11" else
     unaffected;
```

An almost equivalent form is:

```vhdl
z <= "0001" when a = "00" else
     "0010" when a = "01" else
```

```
         "0100" when a = "10" else
         "1000" when a = "11" else
         z;
```

The last version differs from the first two in that any pending assignments to z are thrown away. As we will see in Chapter 6, it cannot be used with the **entity** declaration shown. The interpretation of such statements in hardware terms will be described in detail in Chapter 9 on synthesis, but it is sufficient to note that assigning a signal to itself or not assigning a new value can be interpreted as meaning that a memory element exists. Moreover, as will be described in Chapter 12 on asynchronous design, this memory element is likely to be poorly implemented. Therefore a **when . . . else** statement should normally include the final **else** clause.

The expression in each **when** clause must resolve to a Boolean true or false. In the examples we have simply tested one value of a. We can write more complex logical expressions:

```
z <= "0001" when (a = "00" and (en = '1' or inhibit = '0'))
            else . . .
```

4.2.2   With . . . select statement

An alternative to the **when . . . else** statement is the **with . . . select** statement. Another model of the 2 to 4 decoder is shown below.

```
architecture with_select of decoder is
begin
  with a select
     z <= "0001" when "00",
          "0010" when "01",
          "0100" when "10",
          "1000" when "11",
          "XXXX" when others;
end architecture with_select;
```

At first glance this appears very similar to the **when . . . else** statement, but there are important differences. Each clause of a **when . . . else** statement is interpreted in turn until one expression evaluates to 'true' or, failing that, the final **else** is chosen. In a **with . . . select** statement all the alternatives are checked *simultaneously* to find a matching pattern. Therefore, the **with . . . select** must cover all possible values of the selector. Being of type std_logic_vector, the bits of a can take more than the values '0' and '1', so the **when others** clause must be included. If the **when others** line is omitted, compilation will fail. Equally, the same pattern must not be included in more than one branch. Further, the patterns in the branches must be constants – the patterns must be determined when the VHDL is compiled, not dynamically in the course of a simulation.

If more than one pattern should give the same output, the patterns can be listed. For example, the following model describes a seven-segment decoder that displays the

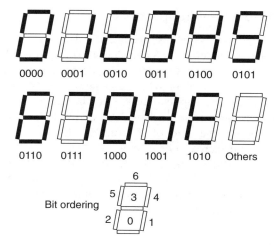

Figure 4.3 Seven-segment display.

digits '0' to '9'. If the bit patterns corresponding to decimal values '10' to '15' are fed into the decoder, an 'E' (for 'Error') is displayed. If the inputs contain X's or other invalid values, the display is blanked. These patterns are shown in Figure 4.3.

```vhdl
library IEEE;
use IEEE.std_logic_1164.all;

entity seven_seg is
  port (a : in std_logic_vector(3 downto 0);
        z : out std_logic_vector(6 downto 0));
end entity seven_seg;

architecture with_select of seven_seg is
begin
  with a select
    z <= "1110111" when "0000",
         "0010010" when "0001",
         "1011101" when "0010",
         "1011011" when "0011",
         "0111010" when "0100",
         "1101011" when "0101",
         "1101111" when "0110",
         "1010010" when "0111",
         "1111111" when "1000",
         "1111011" when "1001",
         "1101101" when "1010"|"1011"|"1100"|
                        "1101"|"1110"|"1111",
         "0000000" when others;
end architecture with_select;
```

4.2.3   *n* to 2^n decoder – shift operators

We have seen two ways to describe a 2 to 4 decoder. The same structures could easily be adapted to model a 3 to 8 decoder or a 4 to 16 decoder. Although these devices are clearly more complex than the 2 to 4 decoder, conceptually there is little difference. It would be convenient to have a general *n* to 2^n decoder that could be described once but used for any application. We saw in the previous chapter that generics can be used to pass parameters, such as delays, to VHDL models. We can similarly use a generic to define the size of a structure. In the entity declaration below, the **generic** n is declared to be of type POSITIVE. POSITIVE is a predefined subtype of INTEGER that can take values in the range 1 to the maximum integer value. If we tried to create a decoder with n equal to 0 or to a negative number, we would get a compilation error. Thus the strong typing of VHDL can be used to ensure we do not get impossible hardware models.

```
library IEEE;
use IEEE.std_logic_1164.all;

entity decoder is
  generic (n : POSITIVE);
  port (a : in std_logic_vector(n-1 downto 0);
        z : out std_logic_vector(2**n-1 downto 0));
end entity decoder;
```

a is now defined to be an *n*-bit vector and z is defined to be a 2^n-bit vector. '******' is the power operator and has a higher precedence than other arithmetic operators (which is why 2**n-1 is interpreted as $2^n - 1$ and not 2^{n-1}). We could have given a default value to n in the generic clause. Whether or not a default value is supplied, n must be defined before the decoder can be simulated or synthesized.

The *n*-bit decoder will have to be written in a different way from the 2-bit decoder. We have noted that the **with . . . select** construct must use constants. We cannot write a list of 2^n constants because we do not know the size of n. Similarly we do not know how many **when . . . else** clauses to write. Looking at the values assigned to z, however, reveals another pattern. The value is always '00 . . . 01' rotated left by the number of places given by the decimal value of a. We can declare a vector of length 2^n with all bits set to '0' other than bit '0' with a constant declaration:

```
constant z_out : std_logic_vector(2**n-1 downto 0) :=
         (0 => '1', others => '0');
```

A constant is declared in the same way as a signal, but (of course) its value can never be changed. The value of the constant is given after the ':=' assignment. In this example, an *aggregate* is used to define the initial value. An aggregate consists of a set of value expressions. In this example, bit 0 is set to '1' and all the others are set to '0'.

VHDL has six shift operators: **sll**, **sla**, **rol**, **srl**, **sra**, and **ror**. The difference between these operators is shown in Figure 4.4.

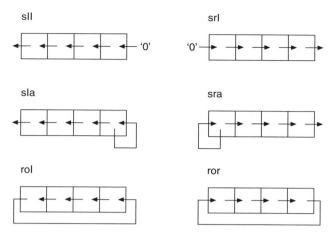

Figure 4.4 VHDL shift operators.

We need to be very careful how we write the code because of VHDL's strong typing. These operators are defined, by default, to shift a BIT_VECTOR by an integer number of places. We want to shift a std_logic_vector by a number of places given by the integer interpretation of another std_logic_vector. Therefore, it would be easier to declare z_out as a BIT_VECTOR, to convert a to an INTEGER and to convert the final result to a std_logic_vector. This last conversion can be done by a function in the std_logic_1164 package. The other conversion function is not, however, provided. To do this we need to use another package, numeric_std, that provides a set of numeric operators for vectors of std_logic – but not std_logic_vectors! Because vectors of bits can be interpreted to be either signed (two's complement) or unsigned integers, we need to distinguish the operations performed on such vectors. Therefore the numeric_std package defines two new types: signed and unsigned. VHDL's strong typing means that we cannot mix signed, unsigned and std_logic_vector by accident, but because all three types consist of arrays of std_logic, we can explicitly convert from one to the other using statements of the kind

```
x <= unsigned(y);

y <= std_logic_vector(x);
```

where x is of type unsigned and y is of type std_logic_vector. Although these look like function calls, no such function has been defined. These are known as *type conversions* (sometimes such a conversion is known as a *cast*). On the other hand, to convert from an unsigned to an INTEGER does require a function call because the possible values ('X', 'Z', etc.) of the std_logic type need to be interpreted. The function to_integer is provided in numeric_std to achieve this. To convert from an INTEGER to an unsigned type, the to_unsigned(i, n) function should be used, where i is the integer and n is the number of bits in the result. The complete model is given below.

```
library IEEE;
use IEEE.std_logic_1164.all;
use IEEE.numeric_std.all;

entity decoder is
  generic (n : POSITIVE);
  port (a : in std_logic_vector(n-1 downto 0);
        z : out std_logic_vector(2**n-1 downto 0));
end entity decoder;

architecture rotate of decoder is
  constant z_out : BIT_VECTOR(2**n-1 downto 0) :=
        (0 => '1', others => '0');
begin
  z <= to_StdLogicVector (z_out sll
                          to_integer(unsigned(a)));
end architecture rotate;
```

4.3 Multiplexers

4.3.1 4 to 1 multiplexer

A multiplexer can be used to switch one of many inputs to a single output. Typically multiplexers are used to allow large, complex pieces of hardware to be reused. The IEEE symbol for a 4 to 1 multiplexer is given in Figure 4.5. G is a select symbol. $\frac{0}{3}$ is not a fraction, but means 0–3. Therefore the binary value on the top two inputs is used to select one of the inputs 0–3.

Two possible models of a 4 to 1 multiplexer are given below.

```
library IEEE;
use IEEE.std_logic_1164.all;

entity mux is
  port (a, b, c, d: in std_logic;
        s: in std_logic_vector(1 downto 0);
```

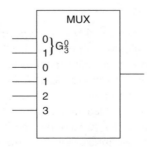

Figure 4.5 4 to 1 multiplexer.

```
         y: out std_logic);
end entity mux;

architecture mux1 of mux is
begin
  with s select
    y <= a when "00",
         b when "01",
         c when "10",
         d when "11",
         'X' when others;
end architecture mux1;

architecture mux2 of mux is
begin
    y <= a when s = "00" else
         b when s = "01" else
         c when s = "10" else
         d when s = "11" else
         'X';
end architecture mux2;
```

Both these forms of multiplexer represent conventional logic. It is also possible to use three-state logic to build a multiplexer. It was noted in Section 4.1 that the outputs of several three-state buffers can be connected together, provided that only one buffer is enabled at one time. A 4 to 1 multiplexer implemented in three-state logic is shown below. There are four assignments to 'y', and therefore four drivers for 'y'. At any time, three are 'Z' and one is an input value. In order for the output value to be correctly determined, and in order not to cause a compilation error, y must be declared to be a resolved type – std_logic. It can be seen from the truth table of the standard logic resolution function that one input is propagated to the output.

```
library IEEE;
use IEEE.std_logic_1164.all;

entity mux is
port (a, b, c, d: in std_logic;
      s: in std_logic_vector(1 downto 0);
      y: out std_logic);
end entity mux;

architecture three_state of mux is
begin
  y <= a when s = "00" else 'Z';
  y <= b when s = "01" else 'Z';
  y <= c when s = "10" else 'Z';
  y <= d when s = "11" else 'Z';
end architecture three_state;
```

4.4 Priority encoder

4.4.1 Don't cares

An encoder takes a number of inputs and encodes them in some way. The difference between a decoder and an encoder is therefore somewhat arbitrary. In general, however, an encoder has fewer outputs than inputs. A priority encoder attaches an order of importance to the inputs. Thus if two inputs are asserted, the most important input takes priority. The symbol for a priority encoder is shown in Figure 4.6. There are three outputs. The lower two are the encoded values of the four inputs. The upper output indicates whether the output combination is valid. An OR function (≥ 1) is used to check that at least one input is 1. Z is used to denote an internal signal. Thus Z10 is connected to 10. This avoids unsightly and confusing lines across the symbol.

An example of a priority encoder is given in the truth table below. The 'Valid' output is used to signify whether at least one input has been asserted and hence whether the outputs are valid.

Inputs				Outputs		
$A3$	$A2$	$A1$	$A0$	$Y1$	$Y0$	Valid
0	0	0	0	0	0	0
0	0	0	1	0	0	1
0	0	1	–	0	1	1
0	1	–	–	1	0	1
1	–	–	–	1	1	1

This piece of VHDL looks as if it ought to model the truth table:

```
library IEEE;
use IEEE.std_logic_1164.all;

entity priority is
  port (a: in std_logic_vector(3 downto 0);
        y: out std_logic_vector(1 downto 0);
        valid: out std_logic);
end entity priority;
```

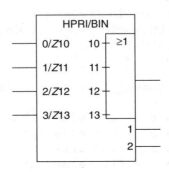

Figure 4.6   4 to 2 priority encoder.

```
architecture DontCare of priority is
begin
  with a select
    y <= "00" when "0001",
         "01" when "001-",
         "10" when "01--",
         "11" when "1---",
         "00" when others;
    valid <= '1' when a(0) = '1' or a(1) = '1' or a(2) = '1'
                   or a(3) = '1' else '0';
end architecture DontCare;
```

However, as far as VHDL is concerned, the 'don't know' logic value is simply another value and cannot be matched to all other possibilities. Therefore this example will not work as intended. One alternative is to select the inputs in order:

```
architecture Ordered of priority is
begin
  y <= "11" when a(3) = '1' else
       "10" when a(2) = '1' else
       "01" when a(1) = '1' else
       "00" when a(0) = '1' else
       "00";
  valid <= '1' when a(0) = '1' or a(1) = '1' or a(2) = '1'
                 or a(3) = '1' else '0';
end architecture Ordered;
```

The numeric_std package includes a function, std_match, that treats the don't care value as a real don't care condition. We can't use std_match in a **with . . . select** statement because the choices must be constant. We can write instead:

```
use IEEE.numeric_std.all;

architecture Match of priority is
begin
  y <= "00" when std_match(a, "0001") else
       "01" when std_match(a, "001-") else
       "10" when std_match(a, "01--") else
       "11" when std_match(a, "1---") else
       "00";
  valid <= '1' when a(0) = '1' or a(1) = '1' or a(2) = '1'
                 or a(3) = '1' else '0';
end architecture Match;
```

(Note that the **use** statement can be prefixed to an architecture, in which case it applies to just that architecture.) This model is awkward, however. There are two signal assignments which duplicate each other to some extent. It would be more natural to test the bits of a and to set y and valid at the same time. To do this in VHDL, we need to use an entirely different style.

4.4.2   Sequential VHDL

There are three styles of VHDL: structural, dataflow and sequential. All the examples in this chapter have used the dataflow style. Dataflow statements are *concurrent signal assignment* statements. Structural VHDL consists of component instantiations. Sequential VHDL resembles a conventional programming language. Sequential VHDL statements can be used only in subprograms (procedures and functions) or processes.

The priority encoder can be described in sequential VHDL as follows:

```vhdl
architecture Sequential of priority is
begin
  process (a) is
  begin
    if a(3) = '1' then
      y <= "11";
      valid <= '1';
    elsif a(2) = '1' then
      y <= "10";
      valid <= '1';
    elsif a(1) = '1' then
      y <= "01";
      valid <= '1';
    elsif a(0) = '1' then
      y <= "00";
      valid <= '1';
    else
      y <= "00";
      valid <= '0';
    end if;
  end process;
end architecture Sequential;
```

Although this is a *sequential* style of programming, this model represents a piece of *combinational* hardware. Models of sequential hardware will be discussed in Chapters 5 and 6. We will also introduce different programming constructs as needed.

The **process** has a *sensitivity list* with one signal, a. The process is evaluated only when the signals in the sensitivity list change. Thus it is important that the sensitivity list includes all signals that might cause an output to change. In this case, a is a vector and the process is evaluated when any bit of a changes.

The main part of the process consists of an **if** statement. In each branch of the **if** statement, assignments are made to y and to valid according to the value of a. In modelling combinational logic, the following rule should be observed: *if an assignment is made to a signal in one path through a process, an assignment* *should be made to that signal in all paths*. If this is not done, the VHDL model will simulate correctly, but a synthesis tool will infer that latches exist in the hardware. This will be discussed further later. It is, however, possible to write the encoder as follows:

```
architecture Sequential2 of priority is
begin
  process (a) is
  begin
    valid <= '1';
    if a(3) = '1' then
      y <= "11";
    elsif a(2) = '1' then
      y <= "10";
    elsif a(1) = '1' then
      y <= "01";
    elsif a(0) = '1' then
      y <= "00";
    else
     valid <= '0';
     y <= "00";
    end if;
  end process;
end architecture Sequential2;
```

The value of `valid` is set at the beginning of the process. If a does not have at least one bit set to '1', `valid` is given a new value in the **else** clause. It might appear that two drivers would be created for `valid`. In fact, this is not the situation. Only one driver per **signal** is created in a **process**. Because a process is sequential, there is no conflict between two assignments to the same signal.

4.5 Adders

4.5.1 Functional model

The IEEE symbol for a 4-bit adder is shown in Figure 4.7. The Σ symbol denotes an adder. *P* and *Q* are assumed to be the inputs to the adder. *CI* and *CO* are carry in and carry out, respectively.

The VHDL entity declaration of an *n*-bit adder can be written as follows:

```
library IEEE;
use IEEE.std_logic_1164.all, IEEE.numeric_std.all;

entity NBitAdder is
  generic (n: NATURAL :=4);
  port (A, B: in std_logic_vector(n-1 downto 0);
        Cin : in std_logic;
```

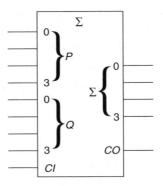

Figure 4.7   4-bit adder.

```
       Sum : out std_logic_vector(n-1 downto 0);
       Cout: out std_logic);
end entity NBitAdder;
```

We have chosen to define the inputs and outputs in terms of std_logic_vectors, but as the numeric_std package has been included, signed or unsigned types could have been used instead. We can use the arithmetic operator, +, defined in numeric_std to perform the adding operation. This operator takes two vectors, of type signed or unsigned, and returns a result *of the same length as the longest operand*. The addition of two n-bit integers produces a result of length $n + 1$, where the most significant bit is the carry out bit. Therefore within the VHDL description we must convert Cin from a single bit to a vector of length $n + 1$, convert A and B to vectors of length $n + 1$ and separate the result into an n-bit sum and a carry out bit. The code below performs these actions for unsigned addition. The ampersand, '&', is the concatenation operator. Thus '0' & unsigned(A) concatenates a single bit and an n-bit vector to give a vector of length $n + 1$. The carry vector is initialized in the same way as the constant declaration in Section 4.2.3. A and B are converted to type unsigned. After addition of three $n + 1$ bit vectors, the lowest n bits of the result are converted back to a std_logic_vector and the most significant bit is taken as the carry out.

```
architecture unsgned of NBitAdder is
  signal result : unsigned(n downto 0);
  signal carry : unsigned(n downto 0);
  constant zeros : unsigned(n-1 downto 0) := (others =>
                                                 '0');
begin
  carry <= (zeros & Cin);
  result <= ('0' & unsigned(A)) + ('0' & unsigned(B))
             + carry;
  Sum <= std_logic_vector(result(n-1 downto 0));
  Cout <= result(n);
end architecture unsgned;
```

The equivalent code for signed arithmetic is given below. The major difference here is that the most significant bits of the A and B vectors are used to extend those vectors to the left. If A or B is negative, its most significant bit would be '1' and this must be preserved.

```
architecture sgned of NBitAdder is
  signal result : signed(n downto 0);
  signal carry : signed(n downto 0);
  constant zeros : signed(n-1 downto 0) := (others => '0');
begin
  carry <= (zeros & Cin);
  result <= (A(n-1) & signed(A)) + (B(n-1) & signed(B))
            + carry;
  Sum <= std_logic_vector(result(n-1 downto 0));
  Cout <= result(n);
end architecture sgned;
```

It should be noted that the underlying structure of these two versions of the adder would be the same. The conversion between types is a feature of VHDL and not of any resulting hardware.

4.5.2   Ripple adder

A simple model of a single-bit full adder might be:

```
library IEEE;
use IEEE.std_logic_1164.all;

entity FullAdder is
  port (a, b, Cin : in std_logic;
        Sum, Cout: out std_logic);
end entity FullAdder;

architecture concurrent of FullAdder is
begin
  Sum <= a xor b xor Cin;
  Cout <= (a and b) or (a and Cin) or (b and Cin);
end architecture concurrent;
```

This model contains two assignments, to Sum and Cout. Note that in VHDL, these two assignments are *concurrent* – it does not matter in which order statements are written. The simple rule to remember is that unless otherwise stated, all statements in VHDL are concurrent.

We can build a multi-bit adder using several instances of this full adder. If we know how many bits will be in our adder we simply instantiate the model several times. If, however, we want to create a general *n*-bit adder, we need some type of iterative construct. The **for . . . generate** construct allows repetition in a dataflow description. This example creates n instances of the FullAdder and, through the `Carry` vector, wires them up. Notice that the loop variable i is implicitly declared.

```
g1 : for i in 0 to n-1 generate
    fm : entity WORK.FullAdder port map (A(i), B(i),
                                        Carry(i), Sum(i),
                                        Carry(i+1));
   end generate g1;
```

We can similarly count down:

```
for i in n-1 downto 0 generate
```

The first and last bits of the adder do not conform to this pattern, however. Bit 0 should have `Cin` as an input and bit $n - 1$ should generate `Cout`. We can make special cases of the first and last elements by using **if . . . generate** statements.

```
architecture StructIterative of NBitAdder is
  signal Carry: std_logic_vector(0 to n);
begin
  g1: for i in 0 to n-1 generate
    lt : if i = 0 generate
      f0 : entity WORK.FullAdder port map
            (A(i), B(i), Cin, Sum(i), Carry(i+1));
    end generate lt;
    rt : if i = n-1 generate
      fn : entity WORK.FullAdder port map
            (A(i), B(i), Carry(i), Sum(i), Cout);
    end generate rt;
    md : if i > 0 and i < n-1 generate
      fm : entity WORK.FullAdder port map
            (A(i), B(i), Carry(i), Sum(i), Carry(i+1));
    end generate md;
  end generate g1;
end architecture StructIterative;
```

Note that there is no 'else' clause to an **if . . . generate**. Note also that all the elements of this description have labels. *Instantiation and generate statements must have labels.*

4.6 Parity checker

The principle of parity checking was explained in Chapter 2. The IEEE symbol for a parity checker is shown in Figure 4.8. The symbol $2k$ indicates that the output is asserted if $2k$ inputs are asserted for any integer k. Thus the output is asserted for even parity. An odd parity checker has the output inverted.

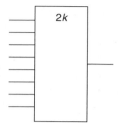

Figure 4.8 Even parity checker.

We could implement a parity checker in VHDL using a network of XOR gates (see Exercises). A more natural method would be to check each bit of the input vector in turn using some kind of loop and to determine the cumulative parity of the bits. VHDL provides a sequential looping construct that has three forms:

loop

while *condition* **loop**

and

for *identifier* **in** *range* **loop**

The first form implements an infinite loop. Although VHDL provides means of breaking out of the loop, this form is probably the least useful of the three. The **while** form is useful when the extent of the loop is not known, but the conditions for exiting the loop are known in advance. The **while** form is not supported by the 1076.6 RTL synthesis standard. The **for** form is probably the most useful. For example, we could loop through all the bits of a vector of length n using a loop of the form:

for i **in** 0 **to** n-1 **loop**

The counting sequence can be reversed by using n-1 **downto** 0. In both cases, the size of the vector, n, would have to be stated explicitly, perhaps using a generic. If the size is not explicitly stated, the loop can be written as:

for i **in** a'RANGE **loop**

where a is the vector. 'RANGE is an *attribute* of a. We will discuss attributes in more detail below. If a has been declared as a vector of range (0 **to** n) then a'RANGE is interpreted as (0 **to** n). Similarly, if a has been declared to have a range (n **downto** 0), a'RANGE is interpreted as (n **downto** 0). The opposite interpretations can be made using the 'REVERSE_RANGE attribute.

It is possible to terminate a loop early using the **exit** statement. Part of a loop can be omitted using a **next** statement to jump from that point to the next loop iteration.

A VHDL model of a parity checker is given below. We want to use the sequential coding constructs **loop** and **if**; therefore a **process** is used. The size of the vector, a, is not stated because the 'RANGE attribute is used as the loop range. To keep track of whether there have been an even or odd number of 1s in the vector, a **variable** is used inside the process. Variables are different from signals in two respects. A variable can only be declared inside a process (a signal may not be declared in a process) and an assignment to a variable (denoted by ':=') takes immediate effect. A signal assignment does not take effect until the process restarts. We will return to this distinction in later chapters. Notice that even is reset to '0' at the top of the process. Variables hold on to their values between activations of a process. If this reset were not done, the result of the last parity evaluation would remain, which could give an incorrect result.

```
library IEEE;
use IEEE.std_logic_1164.all;

entity parity is
  port (a : in std_logic_vector;
        y : out std_logic);
  end entity parity;

architecture iterative of parity is
begin
  process (a) is
    variable even : std_logic;
  begin
    even := '0';
    for i in a'RANGE loop
      if a(i) = '1' then
        even := not even;
      end if;
    end loop;
    y <= even;
  end process;
end architecture iterative;
```

4.6.1   Attributes

Attributes are pieces of information about VHDL units, signals or other types that may be used in models or to control simulators or other tools. For example, the current value of a signal may be passed to another signal by the assignment:

```
a <= b;
```

If we wanted (for whatever reason) to assign the value before the last change of b to a, we could write:

```
a <= b'LAST_VALUE;
```

All attributes are separated from the name to which they refer by an apostrophe (').

We have seen that the range of a vector can be used to control a loop. We could find the vector size explicitly using the attribute 'LENGTH:

```vhdl
entity parity is
  port (a : in std_logic_vector;
        y : out std_logic);
  constant n : NATURAL := a'LENGTH;
end entity parity;
```

It is equally possible to put the **constant** declaration in the **architecture**. Any declaration in an entity is valid for *all* architectures of that entity.

We will meet further predefined attributes later. It is possible to define your own attributes.

4.7 Testbenches for combinational blocks

In the last chapter, we introduced the idea of writing simulation testbenches in VHDL for simple combinational circuits. Testbenches are not synthesizable and therefore the entire scope of VHDL can be used to write them. Testbenches are also notable for the fact that their entity descriptions do not include any ports – a testbench represents the rest of the world.

Two functions are generally performed in a testbench: generation of input stimuli and checking of results. The simple testbenches shown in the last chapter did not perform any checking. Moreover, input stimuli were generated using concurrent assignments. This style is fine for simple circuits, but is not appropriate for circuits with multiple inputs. For example, let us write a testbench for the *n*-bit adder of Section 4.5.1.

```vhdl
library IEEE;
use IEEE.std_logic_1164.all;

entity TestNBitAdder is
end entity TestNBitAdder;

architecture TestBench_1 of TestNBitAdder is
  constant n: NATURAL := 4;
  signal A, B, Sum: std_logic_vector (n-1 downto 0);
  signal Cin, Cout: std_logic;
begin
  s0: entity WORK.NBitAdder(unsgned) generic map (n)
        port map(A, B, Cin, Sum, Cout);
  Cin <= '0', '1' after 10 NS, '0' after 25 NS;
  A <= "0000", "1111" after 5 NS, "0111" after 15 NS;
  B <= "0000", "1111" after 20 NS;
end architecture TestBench_1;
```

There is a very obvious problem with this testbench – it's very difficult to see what is happening and when. It would be a lot clearer if all three signals were updated in one process. We can replace the three concurrent assignment statements with the following process:

```
process is
begin
  Cin <= '0';
  A <= "0000";
  B <= "0000";
  wait for 5 NS;
  A <= "1111";
  wait for 5 NS;
  Cin <= '1';
  wait for 5 NS;
  A <= "0111";
  wait for 5 NS;
  B <= "1111";
  wait for 5 NS;
  Cin <= '0';
  wait;
end process;
```

The behaviour of this process is exactly the same as that of the three assignments. From a human point of view, the big difference is that the time is relative (we wait for 5 ns at a time), rather than absolute. Despite this, it is far easier to see what is happening and, importantly, it is easier to write the testbench and to modify it. Note the inclusion of the final **wait** statement. Without this, the process will simply repeat. This form of the process is different from others that we have seen. In modelling synthesizable hardware, all processes have included a sensitivity list. A process can have a sensitivity list or **wait** statements, but not both (and not neither – a process without any **wait** statements and without a sensitivity list will run forever at time 0).

As far as combinational circuits are concerned, this is about as complex as we ever need to get. It is still difficult, however, to work out what is going on. For example, we try to add '0111' to '0000' with a carry in bit of '1'. The simulation tells us that the sum is '1000' with a carry out bit of '0'. It is just about possible to work out that this is correct, but it is not easy. Instead, we could use integers and convert these to bit patterns.

```
architecture TestBench_3 of TestNBitAdder is
  constant n: NATURAL := 4;
  signal A, B, Sumint : NATURAL;
  signal Aslv, Bslv, Sum: std_logic_vector (n-1 downto 0);
  signal Cin, Cout: std_logic;
begin
s0: entity WORK.NBitAdder(unsgned) generic map (n)
      port map(Aslv, Bslv, Cin, Sum, Cout);
  Aslv <= std_logic_vector(to_unsigned(A, n));
```

```
Bslv <= std_logic_vector(to_unsigned(B, n));
Sumint <= to_integer(unsigned(Cout & Sum));
process is
begin
  Cin <= '0';
  A <= 0;
  B <= 0;
  wait for 5 NS;
  A <= 15;
  wait for 5 NS;
  Cin <= '1';
  wait for 5 NS;
  A <= 7;
  wait for 5 NS;
  B <= 15;
  wait for 5 NS;
  Cin <= '0';
  wait;
 end process;
end architecture TestBench_3;
```

Now we can easily see that $7 + 0 + 1$ is equal to 8 (with no carry out). Better still, we could let the testbench itself check the addition. In general, we do not necessarily want to be told that the design is correct, but we do want to know if there is an error. In Chapter 5, we will see how warning messages can be generated. Another technique is to generate an error signal when unexpected behaviour occurs. It is then relatively easy to spot one signal changing state in a long simulation with lots of complex data.

To the testbench above, we simply add an error signal:

```
signal error: BOOLEAN := FALSE;
```

together with a process that is triggered whenever one of the outputs from the adder changes:

```
resp: process (Cout, Sum) is
      begin
        error <= (A + B + BIT'POS(to_bit(Cin))) /= Sumint;
      end process resp;
```

Notice that we have to convert a single bit of type `std_logic` to an integer. Here, this is done by converting to type `BIT` (values other than '0' or '1' will be converted to '0' using the `to_bit` function) and then using the `'POS` attribute to give the integer position of the value in the type definition. For type `BIT`, '0' is at position 0 and '1' is at position 1.

In later chapters we will see again the principle of performing the operation that we are checking in a different way. We will also use processes, triggered by changing signals, to monitor outputs.

Summary

In this chapter we have introduced a number of typical combinational building blocks. The IEEE standard symbols for these blocks have been described. The VHDL standard logic package has been introduced. Various VHDL constructs have been described: **when . . . else**, **with . . . select**, **generate**, shift operators, the numeric_std package, **process**es, and sequential constructs. In general, the models given in this chapter are suitable for RTL synthesis. Finally, we have seen how testbenches can be constructed to verify VHDL code.

Further reading

A full description of the IEEE symbols is given in the IEEE standard and in a number of digital design textbooks. Manufacturers' data sheets may use the IEEE symbols or a less standard form. VHDL models of synthesizable components (so-called *IP Cores*) can be found by searching the Web. Bergeron covers testbenches in great detail.

Exercises

4.1 What are the three styles of description in VHDL? How are they used? Give a brief example of each.

4.2 VHDL models can be written using *concurrent* and *sequential* coding constructs. Explain, with examples, the meaning of *concurrent* and *sequential* in this context.

4.3 Write an entity description and three architecture models of a 3 to 8 decoder using (a) Boolean operators, (b) a **when . . . else** statement, and (c) a **with . . . select** statement. Write a testbench to compare the three versions concurrently. (Note that you will have to use configuration clauses or direct instantiation to do this.) Simulate the testbench and the decoder models.

4.4 Write a VHDL model of a 2^n to n priority encoder.

4.5 A comparator is used to determine whether two signals have equal values. A one-bit comparator is described by

```
EQO <= EQI and (X xnor Y);
```

where EQI is the result of the comparison of other bits and EQO is passed to the next comparison operation. Write a model of an n-bit iterative comparator.

4.6 Open-drain CMOS logic is used to allow several gates to drive a common bus. Each gate output can be in a 'Z' state or in a '0' state. Only one gate at a time should assert a '0' state. The bus is in a '1' state if all the gates driving it are in a 'Z' state. This is achieved by a pull-up resistor. Design a four-value logic type to model this bus and a suitable resolution function. (The pull-up resistor does not need to be explicitly modelled.) Write a model of an open-drain two-input NAND gate, and hence model a bus driven by four such gates. To simulate this model, the type definitions and the resolution function will need to be put into a package. You may wish to return to this problem after **package**s have been described in Chapter 7.

4.7 What is meant by a *driver* in VHDL? Explain, with an example, how a signal with multiple drivers may be resolved in VHDL.

Synchronous sequential design

We have so far looked at combinational logic design. Real digital systems are usually sequential. Moreover, most sequential systems are synchronous; that is, they are controlled by a clock. In this chapter we will explain how synchronous sequential systems are designed. We will then describe how such systems may be modelled in VHDL.

5.1 Synchronous sequential systems

Almost all large digital systems have some concept of state built into them. In other words, the outputs of a system depend on past values of its inputs as well as the present values. Past input values either are stored explicitly or cause the system to enter a particular state. Such systems are known as *sequential* systems, as opposed to *combinational* systems. A general model of a sequential system is shown in Figure 5.1. The present state of the system is held in the registers – hence the outputs of the registers give the value of the present state and the inputs to the registers will be the next state.

The present state of the system can be updated either as soon as the next state changes, in which case the system is said to be *asynchronous*, or only when a clock signal changes, which is *synchronous* behaviour. In this chapter, we shall describe the

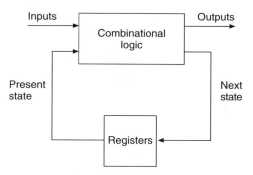

Figure 5.1 General sequential system.

design of synchronous systems. In general, synchronous design is easier than asynchronous design and so we will leave discussion of the latter topic until Chapter 12.

In this chapter we will consider the design of synchronous sequential systems. Many real systems are too complex to design in this way, thus in Chapter 7 we will show that more complex designs can be partitioned. Nevertheless, the formal design methods described in this chapter must be applied to at least part of the design of larger systems. In the next section, we will introduce, by way of a simple example, a method of formally specifying such systems. We will then go on to describe the problems of state assignment, state minimization and the design of the next state and output logic. Throughout we will illustrate how designs can also be modelled using VHDL.

5.2 Models of synchronous sequential systems

5.2.1 Moore and Mealy machines

There are two common models of synchronous sequential systems: the *Moore* machine and the *Mealy* machine. These are illustrated in Figure 5.2. Both types of system are triggered by a single clock. The next state is determined by some (combinational) function of the inputs and the present state. The difference between the two models is that in the Moore machine the outputs are solely a function of the present state, while in the Mealy machine the outputs are a function of the present state and of the inputs. Both the Moore and Mealy machines are commonly referred to as *state machines*. That is to say, they have an internal state that changes.

5.2.2 State registers

As was seen in Chapter 2, combinational logic can contain hazards. The next state logic of the Moore and Mealy machines is simply a block of combinational logic with a number of inputs and a number of outputs. The existence of hazards in this next state logic could cause the system to go to an incorrect state. There are two ways to avoid

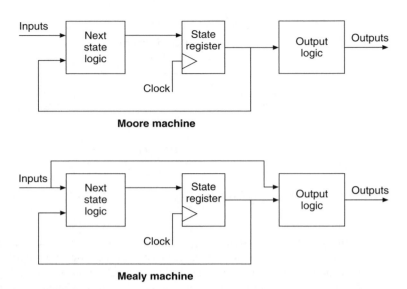

Figure 5.2   Moore and Mealy machines.

such a problem: either the next state logic should include the redundant logic needed to suppress the hazard, or the state machine should be designed such that a hazard is allowed to occur, but is ignored. The first solution is not ideal, as the next state logic is more complex; hence, the second approach is used. (Note that *asynchronous* systems are susceptible to hazards and the next state logic *must* prevent any hazards from occurring, which is one reason why synchronous systems are usually preferred.)

To ensure that sequential systems are able to ignore hazards, a clock is used to synchronize data. When the clock is invalid, any hazards that occur can be ignored. A simple technique, therefore, is to logically AND a clock signal with the system signals – when the clock is at logic 0, any hazards would be ignored. The system is, however, still susceptible to hazards while the clock is high. It is common, therefore, to use registers that are only sensitive to input signals while the clock is changing. The clock edge is very short compared with the period of the clock. Therefore, the data has only to be stable for the duration of the clock change, with small tolerances before and after the clock edge. These timing tolerance parameters are known as the setup and hold times (t_{SETUP}, t_{HOLD}) respectively, as shown in Figure 5.3.

The state registers for a synchronous state machine are therefore edge-triggered elements. The symbol and truth table for a positive edge-triggered D type flip-flop are shown in Figure 5.4. The logic value at the *D* input is stored in the flip-flop, and is available at the *Q* output, at the rising clock edge. In the symbol, the triangle indicates edge-triggered behaviour. A negative edge-triggered flip-flop would have the clock signal inverted (using the usual circle). The notation $C1$, $1D$ shows the dependence of the *D* input on the clock. In the truth table, the notation Q^+ is used to show the next state of *Q* (i.e. after the next clock edge). An upward pointing arrow is used to show a rising edge. Flip-flops may also include asynchronous set or reset inputs, but these should

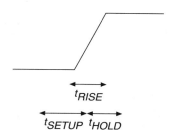

Figure 5.3   Setup and hold times.

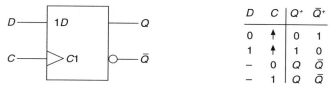

Figure 5.4   D type flip-flop.

only ever be used to initialize the system when it is first turned on. Asynchronous set and reset inputs should *never* be used during normal operation.

Other types of flip-flop exist and may be used to design synchronous systems, but they offer few advantages and are not common in programmable logic. We shall describe these flip-flops in the next chapter.

5.2.3   Design of a three-bit counter

In the next section, we will introduce a formal notation for synchronous sequential systems. First, however, we will consider the design of a simple system that does not need a formal description. Let us design, using positive edge-triggered D flip-flops, a counter that, on rising clock edges, counts through the binary sequence from 000 to 111, at which point it returns to 000 and repeats the sequence.

The three bits will be labelled A, B and C. The truth table is shown below, in which A^+, B^+ and C^+ are the next states of A, B and C.

ABC	$A^+B^+C^+$
0 0 0	0 0 1
0 0 1	0 1 0
0 1 0	0 1 1
0 1 1	1 0 0
1 0 0	1 0 1
1 0 1	1 1 0
1 1 0	1 1 1
1 1 1	0 0 0

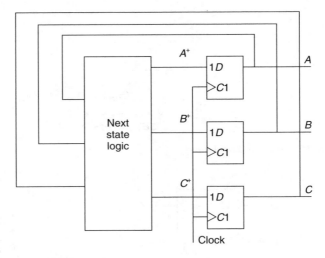

Figure 5.5 Structure of 3-bit counter.

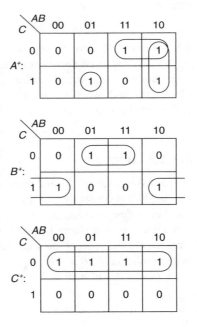

Figure 5.6 K-maps for 3-bit counter.

A^+ etc. are the *inputs* to the state register flip-flops; A etc. are the outputs. Therefore the counter has the structure shown in Figure 5.5. The design task is thus to derive expressions for A^+, B^+ and C^+ in terms of A, B and C. From the truth table above, K-maps can be drawn, as shown in Figure 5.6. Hence the following expressions for the next state variables can be derived.

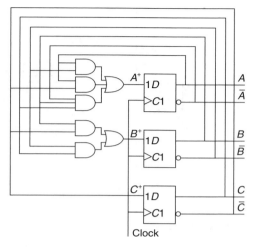

Figure 5.7   3-bit counter circuit.

$$A^+ = A.\overline{C} + A.\overline{B} + \overline{A}.B.C$$
$$B^+ = B.\overline{C} + \overline{B}.C$$
$$C^+ = \overline{C}$$

The full circuit for the counter is shown in Figure 5.7.

5.3   Algorithmic state machines

The counter designed in the last section could easily be described in terms of state changes. Most sequential systems are more complex and require a formal notation to fully describe their functionality. From this formal notation, a state table and hence Boolean expressions can be derived. There are a number of types of formal notation that may be used. We will briefly refer to one before introducing the principal technique used in this book – the *algorithmic state machine (ASM) chart*.

The form of an ASM chart is best introduced by an example. Let us design a simple controller for a set of traffic signals, as shown in Figure 5.8. This example is significantly simpler than a real traffic signal controller (and would probably be more dangerous than an uncontrolled junction!). The traffic signals have two lights each – red and green. The major road normally has a green light, while the minor road has a red light. If a car is detected on the minor road, the signals change to red for the major road and green for the minor road. When the lights change, a timer is started. Once that timer completes, a 'TIMED' signal is asserted, which causes the lights to change back to their default state.

The functionality of this system can be described by the state machine diagram of Figure 5.9. This form of diagram is commonly used, but can be unclear. For some systems (e.g. that of Figure 11.19), such diagrams are sufficient. In this book, however, we

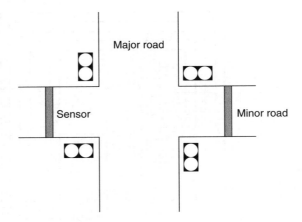

Figure 5.8 Traffic signal problem.

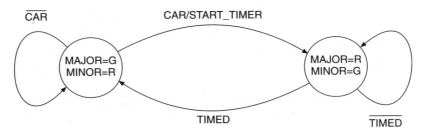

Figure 5.9 State machine of traffic signal controller.

will generally use ASM charts, which are much less ambiguous. The ASM chart for the traffic signal controller is shown in Figure 5.10.

ASM charts resemble flow charts, but contain implicit timing information – the clock signal is not explicitly shown in Figure 5.10. It should be noted that ASM charts represent physical hardware. Therefore all transitions within the ASM chart must form closed paths – hardware cannot suddenly start or stop (the only exception to this might be a reset state to which the system never returns).

The basic component of an ASM chart is the state box, shown in Figure 5.11(a). The state takes exactly one clock cycle to complete. At the top left-hand corner the name of the state is shown. At the top right-hand corner the state assignment (see below) may be given. Within the state box, the output signals are listed. The signals take the values shown for the duration of the clock cycle and are reset to their default values for the next clock cycle. If a signal does not have a value assigned to it (e.g. Y), that signal is asserted (logic 1) during the state and is deasserted elsewhere. The notation $X \leftarrow 1$ means that the signal is assigned at the *end* of the state (i.e. during the next clock cycle) and holds its value until otherwise set elsewhere.

A decision box is shown in Figure 5.11(b). Two or more branches flow from the decision box. The decision is made from the value of one or more input signals. The

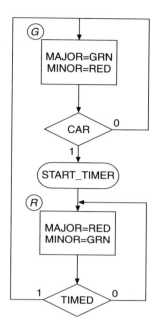

Figure 5.10 ASM chart of traffic signal controller.

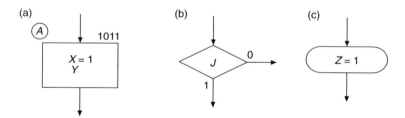

Figure 5.11 ASM chart symbols.

decision box *must* follow and be associated with a state box. Therefore the decision is made in the same clock cycle as the other actions of the state. Hence the input signals must be valid at the start of the clock cycle.

A conditional output box is shown in Figure 5.11(c). A conditional output must follow a decision box. Therefore the output signals in the conditional output box are asserted in the same clock cycle as those in the state box to which it is attached (via one or more decision boxes). The output signals can change during that state as a result of input changes. The conditional output signals are sometimes known as Mealy outputs because they are dependent on input signals, as in a Mealy machine.

It can therefore be seen that one state, or clock cycle, consists of more than just the state box. Decision boxes and conditional output boxes also form part of the state. Figure 5.10 can be redrawn, as in Figure 5.12, where all the components of a state are enclosed within dashed lines.

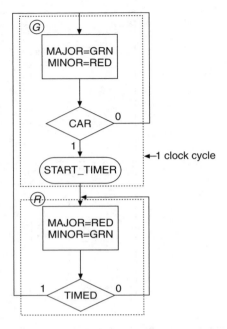

Figure 5.12 ASM chart showing clock cycles.

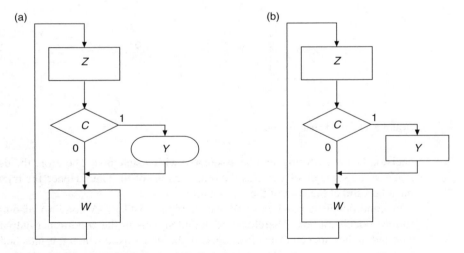

Figure 5.13 Conditional and unconditional outputs.

The difference between state boxes and conditional output boxes is illustrated in Figure 5.13. In Figure 5.13(a), there are two states. Output Y is asserted during the first state if input C is true or becomes true. In Figure 5.13(b) there are three states. The difference can be seen in the timing diagrams of Figure 5.14.

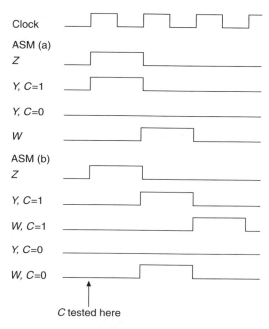

Figure 5.14 Timing diagram for Figure 5.13.

5.4 Synthesis from ASM charts

5.4.1 Hardware implementation

An ASM chart is a description or specification of a synchronous sequential system. It is an abstract description in the sense that it describes *what* a system does, but not *how* it is done. Any given (non-trivial) ASM chart may be implemented in hardware in more than one way. The ASM chart can, however, be used as the starting point of the hardware synthesis process. To demonstrate this, an implementation of the traffic signal controller will first be designed. We will then use further examples to show how the state minimization and state assignment problems may be solved.

The ASM chart of Figure 5.10 may be equivalently expressed as a *state and output table*, as shown in Figure 5.15. The outputs to control the traffic signals themselves are

Present state	CAR, TIMED			
	00	01	11	10
G	G, 0	G, 0	R, 1	R, 1
R	R, 0	G, 0	G, 0	R, 0

Next state, START_TIMER

Figure 5.15 State and output table.

CAR, TIMED

A	00	01	11	10
0	0, 0	0, 0	1, 1	1, 1
1	1, 0	0, 0	0, 0	1, 0

A^+, START_TIMER

Figure 5.16   Transition and output table.

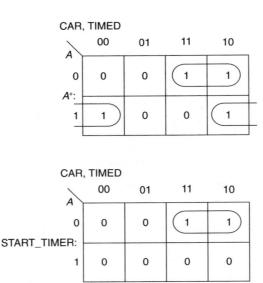

Figure 5.17   K-maps for traffic signal controller.

not shown, but otherwise the state and output table contains the same information as the ASM chart. As we will see, the state and output table is more compact than an ASM chart and is therefore easier to manipulate.

To implement this system in digital hardware, the abstract states G and R have to be represented by Boolean variables. Here, the problem of *state assignment* is nearly trivial. Two states can be represented by one Boolean variable. For example, when the Boolean variable A is 0 it can represent state G, and when it is 1, state R. It would be equally valid to use the opposite values. These values for A can be substituted into the state and output table to give the *transition and output table* shown in Figure 5.16.

This transition and output table is effectively two K-maps superimposed on each other. These are explicitly shown in Figure 5.17. From these, expressions can be derived for the state variable and the output.

$$A^+ = \overline{A}.CAR + A.\overline{TIMED}$$
$$START\_TIMER = \overline{A}.CAR$$

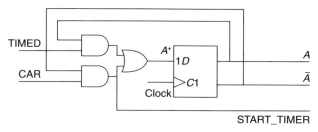

Figure 5.18   Circuit for traffic signal controller.

For completeness, a hardware implementation is shown in Figure 5.18. The two flip-flop outputs can be used directly to control the traffic signals, so that when A is 1 (and $\overline{A}$ is 0) the signal for the major road is green and the signal for the minor road is red. When A is 0, the signals are reversed.

5.4.2   State assignment

In the previous example there were two possible ways to assign the abstract states G and R to the Boolean state variable A. With more states, the number of possible state assignments increases. In general, if we want to code s states using a minimal number of D flip-flops, we need m Boolean variables, where $2^{m-1} < s \leq 2^m$. The number of possible assignments is given by

$$\frac{(2^m)!}{(2^m - s)!}$$

This means, for example, that there are 24 ways to encode three states using two Boolean variables and 6720 ways to encode five states using three Boolean variables. In addition, there are possible state assignments that use more than the minimal number of Boolean variables, which may have advantages under certain circumstances. There is no known method for determining in advance which state assignment is 'best' in the sense of giving the simplest next state logic. It is obviously impractical to attempt every possible state assignment. Therefore a number of *ad hoc* guidelines can be used to perform a state assignment. Again, let us use an example to demonstrate this.

A synchronous sequential system has two inputs, X and Y, and one output, Z. When the sum of the inputs is a multiple of 3, the output is true, otherwise it is false. The ASM chart is shown in Figure 5.19.

To encode the three states we need (at least) two state variables and hence two flip-flops. As noted above, there are 24 ways to encode three states; which should we use? We could arbitrarily choose any one of the possible state assignments, or we could apply one or more of the following guidelines:

● It is good practice to provide some means of initializing the state machine when power is first applied. This can be done using the asynchronous resets or sets on the

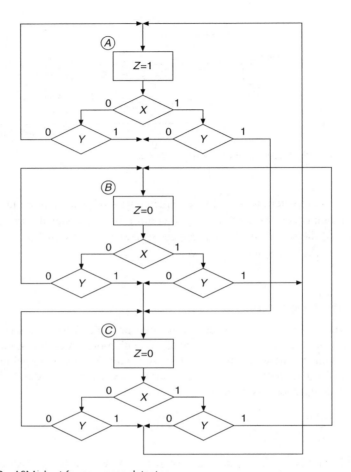

Figure 5.19 ASM chart for sequence detector.

system flip-flops. Therefore the first state (state A in this example) can be coded as all 0s or all 1s.

- We can use the normal binary counting sequence for further states (e.g. B becomes 01 and C becomes 10).
- We can minimize the number of bits that change between states, e.g. by using a Gray code. (This doesn't help in this example as transitions exist from each state to every other state.)
- The states might have some particular meaning. Thus a state variable bit might be set in one state but in no others. (This can result in a non-minimal number of state variables but very simple output logic, which under some circumstances can be very desirable.)
- We can use one variable per state. For three states, we would have three state variables and hence three flip-flops. The states would be encoded as 001, 010 and 100. This is

(a)

	X,Y			
P	00	01	11	10
A	A,1	B,1	C,1	B,1
B	B, 0	C, 0	A, 0	C, 0
C	C, 0	A, 0	B, 0	A, 0

P^+, Z

(b)

	X,Y			
S_1S_0	00	01	11	10
00	00, 1	01, 1	11, 1	01, 1
01	01, 0	11, 0	00, 0	11, 0
11	11, 0	00, 0	01, 0	00, 0

$S_1^+S_0^+, Z$

Figure 5.20   (a) State and output table; (b) transition and output table for sequence detector.

known as 'one-hot' encoding, as only one flip-flop is asserted at a time. Although this appears to be very non-optimal, there may be advantages to the one-hot (or 'one-cold') method. The next state logic may be relatively simple. In some forms of programmable logic, such as FPGAs, there is a very high ratio of flip-flops to combinational logic. A one-hot encoded system may therefore use fewer resources than a system with a minimal number of flip-flops. Furthermore, because exactly one flip-flop output is asserted at a time, it is relatively easy to detect a system malfunction in which this condition is not met. This can be very useful for safety-critical systems.

Let us therefore apply a simple state encoding to the example. The state and output table is shown in Figure 5.20(a) and the transition and output table is shown in Figure 5.20(b), where state A is encoded as 00, B as 01 and C as 11. The combination 10 is not used.

The fact that we have one or more unused combinations of state variables may cause a problem. These unused combinations are states of the system. In normal operation, the system would never enter these 'unused states'. Therefore, in principle, we can treat the next state and output values as 'don't cares', as shown in Figure 5.21.

This gives the next state equations:

$$S_1^+ = S_1.\overline{X}.\overline{Y} + \overline{S_1}.S_0.\overline{X}.Y + \overline{S_0}.X.Y + \overline{S_1}.S_0.X.\overline{Y}$$
$$S_0^+ = S_0.\overline{X}.\overline{Y} + \overline{S_1}.\overline{X}.Y + \overline{S_0}.X + \overline{S_1}.S_0.\overline{Y} + S_1.X.Y$$

The output expression can be read directly from the transition and output table:

$$Z = \overline{S_0}$$

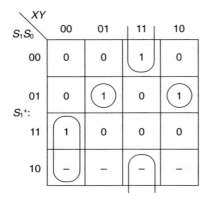 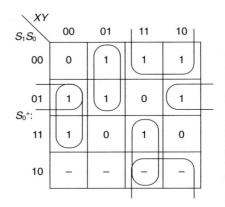

Figure 5.21   K-maps with don't cares.

	X, Y			
$S_1 S_0$	00	01	11	10
00	00, 1	01, 1	11, 1	01, 1
01	01, 0	11, 0	00, 0	11, 0
11	11, 0	00, 0	01, 0	00, 0
10	10, 1	00, 1	11, 1	01, 1

$$S_1^+ S_0^+, Z$$

Figure 5.22   Transition table implied by don't cares.

By default, therefore, the transitions from the unused state have now been defined, as shown in Figure 5.22. Although this unused state should never be entered, it is possible that a 'non-logical' event, such as a glitch on the power supply, might cause the system to enter this unused state. It can be seen from Figure 5.22 that if, for example, the inputs were both 0, the system would stay in the unused state. In the worst case, once having entered an unused state, the system might be stuck in one or more unused states. The unused states could therefore form a 'parasitic' state machine (or perhaps a 'parallel universe'!), causing the system to completely malfunction. We could, reasonably, decide that the chances of entering an unused state are so low as to be not worth worrying about. Hence we treat the transition table entries for the unused states as 'don't cares', as shown, which minimizes the next state logic. On the other hand, the system might be used in a safety-critical application. In this case, it might be important that all transitions from unused states are fully defined, so that we can be certain to return to normal operation as soon as possible. In this case, the transitions from the unused state would not be left as 'don't cares' in the K-maps, but would be explicitly set to lead to, say, the all 0s state. Hence the 'X' entries in the K-maps of Figure 5.21 become 0s and the next state equations would be:

$$S_1^+ = S_1.S_0.\overline{X}.\overline{Y} + \overline{S_1}.S_0.\overline{X}.Y + \overline{S_1}.\overline{S_0}.X.Y + \overline{S_1}.S_0.X.\overline{Y}$$
$$S_0^+ = S_0.\overline{X}.\overline{Y} + \overline{S_1}.\overline{X}.Y + \overline{S_1}.\overline{S_0}.X + \overline{S_1}.\,S_0.\overline{Y} + S_1.S_0.X.Y$$

These equations are more complex than the previous set that includes the 'don't cares'; hence the next state logic would be more complex.

Therefore we have a choice: either we can assume that it is impossible to enter an unused state and minimize the next state equations by assuming the existence of 'don't cares'; or we can try to reduce the risk of becoming stuck in an unused state by explicitly defining the transitions from the unused states and hence have more complex next state logic.

5.4.3   State minimization

We have noted in the previous section that to encode s states we need m flip-flops, where $2^{m-1} < s \leq 2^m$. If we can reduce the number of states in the system, we *might* reduce the number of flip-flops, hence making the system simpler. Such savings may not always be possible. For instance, the encoding of 15 states requires four flip-flops. If we reduced the number of states to nine, we would still need four flip-flops. So there would be no obvious saving and we would have increased the number of unused states, with the potential problems discussed in the previous section. As will be seen, state minimization is a computationally difficult task, and in many cases it would be legitimate to decide that there would be no significant benefits and hence the task would not be worth performing.

State minimization is based on the observation that if two states have the same outputs and the same next states, given a particular sequence of inputs, it is not possible to distinguish between the two states. Hence the two states are considered to be *equivalent* and hence they may be merged, reducing the total number of states.

For example, let us design the controller for a drinks vending machine. A drink costs 40c. The machine accepts 20c and 10c coins (all other coins are rejected by the mechanics of the system). Once 40c have been inserted, the drink is dispensed. If more than 40c are inserted, all coins are returned. The machine has two lights: one to show that it is ready for the next transaction, and one to show that further coins need to be inserted. The ASM chart for the machine is shown in Figure 5.23. The ASM chart has been split into two parts (Figures 5.23(a) and (b)) – the connections between the two parts are shown by circles with lower-case letters.

There are nine states in this state machine. Four flip-flops would therefore be required to implement it. If we could merge at least two states, we would save ourselves a flip-flop. From Figure 5.23 notice that states *F*, *G* and *H* all have transitions to state *I* if a 20c coin is inserted and to state *B* if a 10c coin is inserted. Otherwise all three states have transitions back to themselves. Intuitively, these three states would appear to be equivalent. Another way of looking at this is to say that states *F*, *G* and *H* all represent the condition where another 10c is expected to complete the sale of a drink. From the point of view of the purchaser, these states are indistinguishable.

(a)

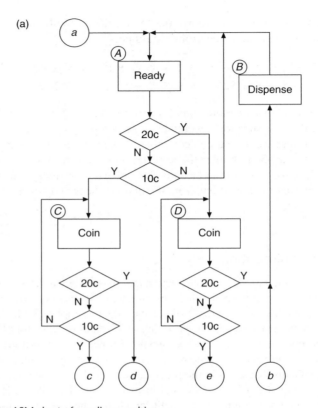

Figure 5.23(a)   ASM chart of vending machine.

Instead of attempting to manipulate the ASM chart, it is probably clearer to rewrite it as a state and output table (Figure 5.24). The 'Other' column shows the next state if no valid coin is inserted. Because there are no conditional outputs, it is possible to separate the outputs from the next state values.

The condition for two states to be considered equivalent is that their next states and outputs should be the same. States A, B and I have unique outputs and therefore cannot be equivalent to any other states. States C to H inclusive have the same outputs. States F, G and H have the same next states, other than their default next states, which are the states themselves. In other words, states F, G and H are equivalent if states F, G and H are equivalent – which is a tautology! Therefore we can merge these three states. In other words, we will delete states G and H, say, and replace all instances of those two states with state F (Figure 5.25). Now states D and E are equivalent, so E can be deleted and replaced by D (Figure 5.26). The system has therefore been simplified from having nine states to having six. It should be remembered that the system may be implemented with nine states or with six, but it is not possible for an external observer to know which version has been built simply by observing the outputs. The two versions are therefore functionally identical.

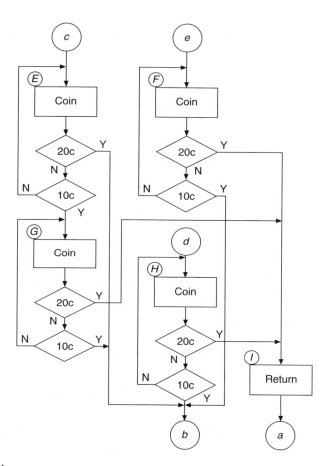

Figure 5.23(b)

State	20c	10c	Other	Ready	Dispense	Return	Coin
A	D	C	A	1	0	0	0
B	A	A	A	0	1	0	0
C	H	E	C	0	0	0	1
D	B	F	D	0	0	0	1
E	B	G	E	0	0	0	1
F	I	B	F	0	0	0	1
G	I	B	G	0	0	0	1
H	I	B	H	0	0	0	1
I	A	A	A	0	0	1	0

Figure 5.24   State and output table for vending machine.

State	20c	10c	Other	Ready	Dispense	Return	Coin
A	D	C	A	1	0	0	0
B	A	A	A	0	1	0	0
C	~~H~~ F	E	C	0	0	0	1
D	B	F	D	0	0	0	1
E	B	~~G~~ F	E	0	0	0	1
F	I	B	F	0	0	0	1
~~G~~	~~I~~	~~B~~	~~G~~	~~0~~	~~0~~	~~0~~	~~1~~
~~H~~	~~I~~	~~B~~	~~H~~	~~0~~	~~0~~	~~0~~	~~1~~
I	A	A	A	0	0	1	0
	Next state			Outputs			

Figure 5.25   State table with states G and H removed.

State	20c	10c	Other	Ready	Dispense	Return	Coin
A	D	C	A	1	0	0	0
B	A	A	A	0	1	0	0
C	~~H~~ F	~~E~~ D	C	0	0	0	1
D	B	F	D	0	0	0	1
~~E~~	~~B~~	~~G~~ F	~~E~~	~~0~~	~~0~~	~~0~~	~~1~~
F	I	B	F	0	0	0	1
~~G~~	~~I~~	~~B~~	~~G~~	~~0~~	~~0~~	~~0~~	~~1~~
~~H~~	~~I~~	~~B~~	~~H~~	~~0~~	~~0~~	~~0~~	~~1~~
I	A	A	A	0	0	1	0
	Next state			Outputs			

Figure 5.26   State table with states E, G and H removed.

To conclude this example, the next state and output expressions will be written assuming a 'one-hot' implementation, i.e. there is one flip-flop per state, of which exactly one has a '1' output at any time. The next state and output expressions can be read directly from the state and output table of Figure 5.26.

$$A^+ = B + I + \overline{20c}.\overline{10c}.A$$
$$B^+ = D.20c + F.10c$$
$$C^+ = A.10c + \overline{20c}.\overline{10c}.C$$
$$D^+ = A.20c + C.10c + \overline{20c}.\overline{10c}.D$$
$$F^+ = C.20c + D.10c + \overline{20c}.\overline{10c}.F$$
$$I^+ = F.20c$$
$$\text{Ready} = A$$
$$\text{Dispense} = B$$
$$\text{Return} = I$$
$$\text{Coin} = C + D + F$$

5.5   State machines in VHDL

5.5.1   A first example

Although state machines can be described using concurrent VHDL constructs, the task is far easier using sequential VHDL. We have seen that a VHDL process is evaluated when a signal in its sensitivity list changes. If the process models a combinational block, all the combinational inputs must be present in the sensitivity list to cause the process to be evaluated. A process may alternatively contain one or more **wait** statements. A process cannot have both a sensitivity list and **wait** statements. A sensitivity list is equivalent to putting a **wait** statement with the signals listed at the end of the process. A state machine changes state at a clock edge. Therefore, the sensitivity list of a process modelling a state machine must include the clock, or the clock must be included in a **wait** statement. A decision to change state then has to be made on the appropriate clock edge.

The state of the system must be held in an internal variable. The state can be represented by an *enumerated type*. The possible values of this type are the state names, e.g.

```
type state_type is (G, R);
```

Because the abstract state names are used, it is not necessary to perform a state assignment. Moreover, because only valid states are listed, there is automatic range checking. This would not be the case if integers or Boolean values were used.

In the following listing, a **variable** is used to hold the current state value. A **case** statement is used to branch according to the current value of the state. Each branch of the **case** statement is therefore equivalent to one of the 'large states' of Figure 5.12. Within the first **when** statement branch, the two lights are set – when major_green is 1, the signal on the major road is green; otherwise it is red. The same interpretation applies to minor_green. The process then waits until the clock input changes to '1'. The meaning of the **wait until** statement is sometimes misunderstood. If the clock is already at '1', the process waits until the clock changes to '0' and back to '1'. Therefore this statement causes the process to wait for a rising edge. Once a rising clock edge has occurred, the car input is tested to set the start_timer output. Note that start_timer is given a default value at the beginning of the process. This is good practice, as it ensures that latches will not be accidentally created if the state machine model is used with a synthesis tool (see Chapter 9).

The other state is structured in a similar way.

```
library IEEE;
use IEEE.std_logic_1164.all;

entity traffic is
  port (clock, timed, car : in std_logic;
        start_timer, major_green,
        minor_green : out std_logic);
end entity traffic;
```

```vhdl
architecture asm1 of traffic is
begin
  process is
    type state_type is (G, R);
    variable state : state_type;
  begin
      start_timer <= '0';
      case state is
        when G =>
          major_green <= '1';
          minor_green <= '0';
          wait until clock = '1';
          if (car = '1') then
              start_timer <= '1';
              state := R;
          end if;
        when R =>
          major_green <= '0';
          minor_green <= '1';
          wait until clock = '1';
          if (timed = '1') then
              state := G;
          end if;
      end case;
  end process;
end architecture asm1;
```

In fact, this is an unusual way to write a state machine model. It's not wrong – a simulation would demonstrate that the state machine model behaves as expected, but the structure would not be recognized by many synthesis tools. It is more usual to have one edge detection statement at the beginning of a process. We could write the process as follows:

```vhdl
process (clock) is
  type state_type is (G, R);
  variable state : state_type;
begin
  start_timer <= '0';
  wait until clock = '1'
  case state is
    when G =>
      major_green <= '1';
      minor_green <= '0';
      if (car = '1') then
        start_timer <= '1';
        state := R;
      end if;
```

```
    when R =>
      major_green <= '0';
      minor_green <= '1';
      if (timed = '1') then
         state := G;
      end if;
  end case;
end process;
```

This is almost but not quite the same as the first version. The difference is in the behaviour of the `start_timer` output. A Mealy output depends on the present state and the values of any input signals. In the example shown, the output will change only at a clock edge. As most synthesis tools and the 1076.6 RTL Synthesis standard expect there to be one edge-sensitive statement in a process, it is not possible to correctly model state machines with Mealy outputs using a single VHDL process.

A common modelling style for state machines therefore uses two processes. One process is used to model the state registers, while the second process models the next state and output logic. The two processes therefore correspond to the two boxes in Figure 5.1. From Figure 5.1, it can be seen that the communication between the two processes is achieved using the present and next values of the state registers. Therefore if two VHDL processes are used, communication between them must be performed using present and next state **signal**s. In VHDL, a **variable** (other than a **shared variable**) can exist only within a process. Therefore, the **type** definition and the **signal** declarations have to occur at the head of the **architecture**.

The two processes shown below have been given labels: `seq` and `com`. These labels are also included in the **end process** lines. For the rest of this chapter, we will use a sensitivity list with the clock signal instead of a **wait** statement. The process is triggered on both clock edges, so we use the `rising_edge` function to distinguish between them. When a rising edge is detected, the value of the `next_state` signal is assigned to the `present_state` signal.

The `com` process has the two inputs, `car` and `timed`, in its sensitivity list, together with the `present_state` signal. The remainder of the process is similar to the previous versions except that the **case** statement selects on the `present_state`, and `next_state` is updated. Note also that `next_state` is updated (to its existing value) even when a change of state does not occur. Failure to do this would result in unnecessary latches being created if the state machine model were fed through a synthesis tool. To repeat the observation made in Section 4.4.2, a process that models combinational logic must assign a value to a signal in *every* path through that process if an assignment is made to that signal in *any* path.

```
architecture asm2 of traffic is
  type state_type is (G, R);
  signal present_state, next_state : state_type;
begin
  seq: process (clock) is
```

```
begin
  if (rising_edge(clock)) then
    present_state <= next_state;
  end if;
end process seq;
com: process (car, timed, present_state) is
begin
  start_timer <= '0';
  case present_state is
    when G =>
      major_green <= '1';
      minor_green <= '0';
      if (car = '1') then
        start_timer <= '1';
        next_state <= R;
      else
        next_state <= G;
      end if;
    when R =>
      major_green <= '0';
      minor_green <= '1';
      if (timed = '1') then
        next_state <= G;
      else
        next_state <= R;
      end if;
  end case;
end process com;
end architecture asm2;
```

The two-process model, registers and next state and output logic, is a very common way to model state machines in VHDL. It is also possible to have a three-process model. The register process is the same as before, but the next state (ns) and output (op) logic blocks are separated, as in the Moore and Mealy models (Figure 5.2). For example:

```
ns: process (car, timed, present_state) is
begin
  case present_state is
    when G =>
      if (car = '1') then
        next_state <= R;
      else
        next_state <= G;
      end if;
    when R =>
      if (timed = '1') then
```

```
            next_state <= G;
         else
            next_state <= R;
         end if;
   end case;
end process ns;

op: process (car, present_state) is
begin
   start_timer <= '0';
   if (present_state = G) then
      major_green <= '1';
      minor_green <= '0';
      if (car = '1') then
         start_timer <= '1';
      end if;
   else
      major_green <= '0';
      minor_green <= '1';
   end if;
end process op;
```

The op process could also be written as concurrent statements:

```
start_timer <= '1' when (present_state = G and car = '1')
                    else '0';
major_green <= '1' when (present_state = G) else '0';
minor_green <= '1' when (present_state = R) else '0';
```

It does not matter which style (one, two or three processes) is used to model a state machine; for different applications one style may be more appropriate than another. If the model is to be synthesized, however, care must be taken to ensure that processes modelling combinational logic really are combinational and that processes modelling sequential logic have only one edge-sensitive statement.

5.5.2 A sequential parity detector

Consider the following system. Data arrives at a single input, with one new bit per clock cycle. The data is grouped into packets of four bits, where the fourth bit is a parity bit. (This problem could easily be scaled to have more realistically sized packets.) The system uses even parity. In other words, if there is an odd number of 1s in the first three bits, the fourth bit is a 1. If an incorrect parity bit is detected, an error signal is asserted during the fourth clock cycle.

The parity detector can be implemented as a state machine. We will leave the design as an exercise and simply show a VHDL implementation. In this example, an asynchronous reset is included to set the initial state to s0. Notice that the error signal is given a default value at the top of process com.

```vhdl
library IEEE;
use IEEE.std_logic_1164.all;

entity parity is
  port (clock, reset, a : in std_logic;
        error : out std_logic);
  end entity parity;

architecture asm of parity is
  type state is (s0, s1, s2, s3, s4, s5, s6);
  signal present_state, next_state : state;
begin
  seq: process (reset, clock) is
  begin
    if reset = '1' then
      present_state <= s0;
    elsif rising_edge(clock) then
      present_state <= next_state;
    end if;
  end process seq;
  com: process (present_state, a) is
  begin
    error <= '0';
    case present_state is
      when s0 =>
        if a = '0' then
          next_state <= s1;
        else
          next_state <= s2;
        end if;
      when s1 =>
        if a = '0' then
          next_state <= s3;
        else
          next_state <= s4;
        end if;
      when s2 =>
        if a = '0' then
          next_state <= s4;
        else
          next_state <= s3;
        end if;
      when s3 =>
        if a = '0' then
          next_state <= s5;
        else
          next_state <= s6;
```

```
        end if;
      when s4 =>
        if a = '0' then
          next_state <= s6;
        else
          next_state <= s5;
        end if;
      when s5 =>
        if a = '0' then
          next_state <= s0;
        else
          error <= '1';
          next_state <= s0;
        end if;
      when s6 =>
        if a = '0' then
          error <= '1';
          next_state <= s0;
        else
          next_state <= s0;
        end if;
    end case;
  end process com;
end architecture asm;
```

5.5.3   Vending machine

The following piece of VHDL is a model of the (minimized) vending machine of
Section 5.4.3. Two processes are used. Note that here an asynchronous reset has been
provided to initialize the system when it is first turned on.

```
library IEEE;
use IEEE.std_logic_1164.all;

entity vending is
  port (clock, reset, twenty, ten : in std_logic;
        ready, dispense, ret, coin : out std_logic);
end entity vending;

architecture asm of vending is
  type state_type is (A, B, C, D, F, I);
  signal present_state, next_state : state_type;
begin
  seq: process (clock, reset) is
  begin
    if (reset = '1') then
      present_state <= A;
```

```vhdl
      elsif (rising_edge(clock)) then
        present_state <= next_state;
      end if;
  end process seq;
com: process (twenty, ten, present_state) is
begin
  ready <= '0';
  dispense <= '0';
  ret <= '0';
  coin <= '0';
  case present_state is
    when A =>
      ready <= '1';
      if (twenty = '1') then
        next_state <= D;
      elsif (ten = '1') then
        next_state <= C;
      else
        next_state <= A;
      end if;
    when B =>
      dispense <= '1';
      next_state <= A;
    when C =>
      coin <= '1';
      if (twenty ='1') then
        next_state <= F;
      elsif (ten ='1') then
        next_state <= D;
      else
        next_state <= C;
      end if;
    when D =>
      coin <= '1';
      if (twenty = '1') then
        next_state <= B;
      elsif (ten = '1') then
        next_state <= F;
      else
        next_state <= D;
      end if;
    when F =>
      coin <= '1';
      if (twenty = '1') then
        next_state <= I;
```

```
        elsif (ten = '1') then
          next_state <= B;
        else
          next_state <= F;
        end if;
      when I =>
        ret <= '1';
        next_state <= A;
    end case;
  end process com;
end architecture asm;
```

5.5.4   Storing data

One of the many problems with the traffic light controller of Section 5.5.1 is that the minor road lights will switch to green as soon as a car is detected. This will happen even if the lights have just changed. It would be preferable if the timer were used to keep the major road lights at green for a period of time. If we did this simply by asserting the start_timer signal in both states and waiting for the timed signal to appear, as follows, an arriving car could easily be missed.

```
com: process (car, timed, present_state) is
begin
  start_timer <= '0';
  case present_state is
    when G =>
      major_green <= '1';
      minor_green <= '0';
      if (car = '1' and timed = '1') then
        start_timer <= '1';
        next_state <= R;
      else
        next_state <= G;
      end if;
    when R =>
      major_green <= '0';
      minor_green <= '1';
      if (timed = '1') then
        start_timer <= '1';
        next_state <= G;
      else
        next_state <= R;
      end if;
  end case;
end process com;
```

Therefore, the fact that a car has arrived needs to be remembered in some way. This could be done by adding further states to the state machine. Alternatively, the car arrival could be stored. It is not possible to say that one approach is better than the other. We will look at the idea of using a state machine to control other hardware in Chapter 7. Meanwhile, let us consider how a simple piece of data can be stored.

In a purely simulation model, it is possible to store the state of a variable or signal in a combinational process. This is done by assigning a value in one branch of the process. As we will see in Chapter 9, when synthesized this would inevitably lead to asynchronous latches and hence timing problems. Instead, any data that is to be stored must be explicitly saved in a register, modelled as a clocked process. Storing data in this way is exactly the same as storing a state. Therefore, separate signals are needed for the present value of the car register and for the next value. We will use more meaningful names for these signals:

```
signal car_arrived, car_waiting : std_logic;
```

The `car_waiting` signal is updated at the same time as the `present_state` signal.

```
seq: process (clock) is
  begin
    if rising_edge(clock) then
      present_state <= next_state;
      car_waiting <= car_arrived;
    end if;
  end process seq;
```

The `car_arrived` signal is set or reset in the following process:

```
car_update: process (car, car_waiting, timed,
                     present_state) is
  begin
    if (present_state = G and car_waiting = '1'
        and timed = '1') then
      car_arrived <= '0';
    elsif car = '1' then
      car_arrived <= '1';
    else
      car_arrived <= car_waiting;
    end if;
  end process car_update;
```

Finally, both references to `car` in process `com` at the start of this section need to be replaced by references to `car_waiting`. Notice that each signal is assigned in only one process. It often helps to sketch a diagram of the system with each process represented by a box and showing all the inputs and outputs of each process. If a signal appears to be an output from two boxes, or if a signal is not an input to a process, something is not right!

5.6 VHDL testbenches for state machines

In the last chapter, we looked at how testbenches for combinational circuits can be designed. Here and in the next chapter, we will consider testbenches for sequential circuits. In this section we will consider clock generation, modelling asynchronous resets and other deterministic signals and how to synchronize inputs with the clock. We will also look at collecting responses. In the next chapter, we will extend these ideas to include a degree of randomness in signals.

5.6.1 Clock generation

The most important signal in any design is the clock. In the simplest case, a clock can be generated by inverting its value at a regular interval:

```
clock <= not clock after 10 NS;
```

If clock is of type BIT, the initial value is automatically '0'. If clock is of type std_logic, an initial value must be assigned at the time of declaration:

```
signal clock : std_logic := '0';
```

If this initialization is not performed, the initial value is 'U', and '**not** 'U'' is also 'U'.

The same effect can be achieved by explicitly assigning values to the clock within a process:

```
clk: process is
  begin
    clock <= '0';
    wait for 10 NS;
    clock <= '1';
    wait for 10 NS;
  end process clk;
```

5.6.2 Reset and other deterministic signals

After the clock, the next most important signal is probably the reset (or set). The clock generation process repeats, but the reset signal is usually asserted only once, at the beginning of a simulation, so the process is prevented from repeating by putting an unconditional **wait** statement at the end.

```
rst: process is
  begin
    reset <= '1';
    wait for 5 NS;
    reset <= '0';
    wait for 5 NS;
    reset <= '1';
    wait;
  end process rst;
```

This is exactly the same, in form, as the signal generation process for combinational circuits as given in the last chapter. Note that the reset is deasserted at the start of the simulation and asserted a short time later. This is to ensure that the state of the circuit prior to the reset can be observed.

In exactly the same way, other inputs to a state machine can be generated. The difficulty with this is that it soon becomes easy to lose track of how many clock cycles have passed or whether a signal is asserted in time for a rising edge or for a falling edge. For testing many types of state machine and other sequential circuit, it may be desirable to synchronize input changes with the clock.

5.6.3   Synchronized inputs

In exactly the same way that an RTL model can be made sensitive to the clock or to some other signal, parts of the testbench can also be made sensitive to the clock. For clarity, however, we may wish to delay a signal slightly. This suggests the use of a **wait for** statement, but we cannot mix **wait** statements and sensitivity lists.

```
inc: process is
  begin
    wait until clock = '1';
    wait for 5 NS;
    a <= a + 1;
  end process inc;
```

The **wait until** clock = '1' statement means wait until the clock changes to 1, i.e. wait until the next rising edge. If the clock is already 1, we still wait for the next rising edge. After that edge, the process then waits for a further 5 ns, before performing some data operation. The process repeats each clock cycle (provided the clock period is greater than 5 ns). Therefore an input can be updated without having to count clock cycles.

Similarly, we can synchronize with an output signal. In the traffic light controller example, we can emulate the timer with the following process.

```
tim: process is
  begin
    wait until start_timer = '1';
    timed <= '0';
```

```
   wait for 100 NS;
   timed <= '1';
end process tim;
```

5.6.4   Checking responses

In the last chapter, we saw how an error signal can be generated if a combinational model behaves in a different way from that expected. For a state machine, we can monitor output changes in a similar way. Here, we will output a message when an output changes.

```
mon: process(major_green, minor_green) is
        variable lb : line;
     begin
        write(lb, NOW);
        if major_green = '1' then
          write(lb, STRING'(" Major Road is Green"));
        elsif minor_green = '1' then
          write(lb, STRING'(" Minor Road is Green"));
        end if;
        writeline(output, lb);
     end process mon;
```

This example uses the built-in text output routines. In order to make these visible the following clause must be inserted before the entity declaration of the testbench:

```
use STD.textio.all;
```

Two forms of the `write` procedure are shown. Both copy a string to an output line buffer, `lb`. First of all the current simulation time, given by the function `NOW`, is copied to `lb`. Then a string is written to `lb`, depending on which of the outputs is asserted. The `STRING` qualifier is needed to ensure that the correct form of the `write` procedure is chosen. Finally, the line buffer is copied to the `output` with `writeline`. This might appear to be a complicated way to construct a monitor process, but it should be noted that an `std_logic` signal cannot be passed to a `write` procedure using standard packages.

Summary

State machines can be formally described using ASM charts. The design of a synchronous state machine from an ASM chart has a number of distinct steps: state minimization, state assignment, derivation of next state, and output logic. By using abstract data types, a VHDL model of a state machine can be written that is equivalent to an ASM chart. This VHDL model may be automatically synthesized to hardware using an RTL synthesis tool.

Further reading

State machine design is a core topic in digital design and therefore covered in many textbooks. Not all books use ASM notation; many use the style of Figure 5.9. The problem of state minimization is covered in detail in books such as Hill and Peterson.

5.1 Explain the difference between a Mealy machine and a Moore machine.

5.2 Describe the symbols used in an ASM diagram.

5.3 The following code shows part of a VHDL description of a synchronous state machine. Complete the description by writing down the synchronization process. How would an asynchronous reset be included?

```vhdl
entity state_machine is
    port( x, clock :in BIT;
          z :out BIT);
end entity state_machine;

architecture behaviour of state_machine is
  type state_type is (S0, S1, S2, S3);
  signal state, next_state : state_type;
-- synchronization statements go here!
com: process (state, X) is
  begin
    case state is
      when S0 =>
        Z <= '0';
        if X = '0' then
          next_state <= S0;
        else
          next_state <= S2;
        end if;
      when S1 =>
        Z <= '1';
        if X = '0' then
          next_state <= S0;
        else
          next_state <= S2;
        end if;
      when S2 =>
        Z <= '0';
        if X = '0' then
          next_state <= S2;
```

```
        else
            next_state <= S3;
        end if;
      when S3 =>
        Z <= '0';
        if X = '0' then
            next_state <= S3;
        else
            next_state <= S1;
        end if;
    end case;
  end process com;
end architecture behaviour;
```

5.4 Draw the ASM chart that describes the state machine shown in Exercise 5.3.

5.5 Draw an ASM chart to describe a state machine that detects a sequence of three logical 1s occurring at the input and that asserts a logical 1 at the output *during* the last state of the sequence. For example, the sequence 001011101111 would produce an output 000000100011. Write a two-process VHDL description of the state machine.

5.6 Write a testbench to simulate the state machine of Exercise 5.5 and verify the VHDL model by simulation.

5.7 Produce next state and output logic for the state machine of Exercise 5.5 and write a VHDL description of the hardware using simple gates and positive edge-triggered D flip-flops. Verify this hardware by simulation.

5.8 A state machine has two inputs, A and B, and one output, Z. If the sequence of input pairs $A = 1$ $B = 1$, $A = 1$ $B = 0$, $A = 0$ $B = 0$ is detected, Z becomes 1 during the final cycle of the sequence, otherwise the output remains at 0. Write a two-process VHDL model of a state machine to implement this system.

5.9 Rewrite the model of Exercise 5.8 to use three processes (or concurrent statements): one for the registers, one for the next state logic and one for the output logic.

5.10 Rewrite the model of Exercise 5.8 to use only one process.

5.11 Design, using an ASM chart, a traffic signal controller for a crossroads. The signals change only when a car is detected in the direction with a red signal. The signals change in the (British) sequence: Red, Red and Amber, Green, Amber, Red. Note that while the signals in one direction are Green, Amber, or Red and Amber, the signals in the other direction are Red (i.e. you need more than four states). Design an implementation that uses a minimal number of D flip-flops.

5.12 A counter is required to count people entering and leaving a room. The room has a separate entrance and exit. Sensors detect people entering and leaving. Up to seven people are allowed in the room at one time. Draw an ASM chart of a synchronous counter that counts the people in the room and that indicates when the room is empty and full. One person may enter and one person may leave during each clock cycle. The empty and full indicators should be asserted immediately the condition is true, i.e. before the next clock edge. Write a VHDL model of the system.

5.13 Construct a state and output table for the state machine represented by Figure 5.27. Show that the number of states can be reduced. Derive the next state and output logic to implement the reduced state machine using (a) a minimal number of D flip-flops, and (b) the 'one-hot' D flip-flop method. What are the relative advantages of each method? How has the reduction in the number of states helped in each case?

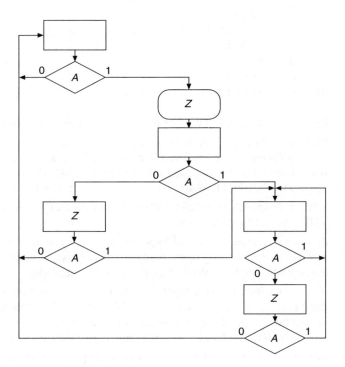

Figure 5.27   ASM chart for Exercise 5.13.

Chapter 6

VHDL models of sequential logic blocks

In Chapter 4 we presented several examples of combinational building blocks, at the same time introducing various aspects of VHDL. In this chapter we shall repeat the exercise for sequential blocks.

6.1 Latches

6.1.1 SR latch

There is often confusion between the terms 'latch' and 'flip-flop'. Here, we will use 'latch' to mean a level-sensitive memory device and 'flip-flop' to specify an edge-triggered memory element. We will discuss the design of latches and flip-flops in Chapter 12. We will simply note here that a latch is based on cross-coupled gates, as shown in Figure 6.1. Table 6.1 gives the truth table of this latch.

When S and R are both at logical 1, the latch holds on to its previous value. When both are at 0, both outputs are at 1. It is this latter behaviour that makes the SR latch unsuitable for designing larger circuits, as a latch or flip-flop would normally be expected to

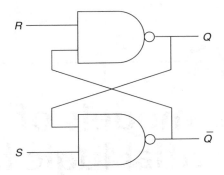

Figure 6.1   SR latch.

Table 6.1   Truth table of SR latch.

S	R	Q	$\overline{Q}$
0	0	1	1
0	1	0	1
1	0	1	0
1	1	Q	$\overline{Q}$

have different values at its two outputs, and it is difficult to ensure that both inputs will never be 0 at the same time.

The SR latch could be modelled in VHDL in a number of ways. Two examples are shown below.

```
library IEEE;
use IEEE.std_logic_1164.all;

entity SR_latch1 is
  port (S, R : in std_logic;
        Q, Qbar : buffer std_logic);
end entity SR_latch1;

architecture dataflow of SR_latch1 is
begin
Q <= '1' when R = '0' else
     '0' when S = '0' else
       Q;
Qbar <= '1' when S = '0' else
        '0' when R = '0' else
          Qbar;
end architecture dataflow;

library IEEE;
use IEEE.std_logic_1164.all;
```

```
entity SR_latch2 is
  port (S, R : in std_logic;
        Q, Qbar : out std_logic);
end entity SR_latch2;

architecture behavioural of SR_latch2 is
begin
p0: process (R, S) is
  begin
    case std_logic_vector'(R, S) is
      when "00" =>
        Q <= '1';
        Qbar <= '1';
      when "01" =>
        Q <= '1';
        Qbar <= '0';
      when "10" =>
        Q <= '0';
        Qbar <= '1';
      when others =>
        null;
    end case;
  end process p0;
end architecture behavioural;
```

In the first architecture, Q and Qbar are both written to and read in the final **else** clauses. As was suggested in Section 4.2.1, this is not permitted if the ports are declared with mode **out**. Therefore, Q and Qbar are declared to have mode **buffer**. They could alternatively have been declared to have mode **inout**. An **inout** mode port is a true bidirectional port. A **buffer** port, on the other hand, is an output port that can be read but which must have only a single driver. As the name suggests, the signal is modelled as if the internal signal is connected to the external port through a buffer. There is, however, a catch. In the 1993 standard, when an entity with a port of mode **buffer** is used inside another entity, that port may only be connected to an internal signal or to another port of mode **buffer**. Therefore the two models shown may not be interchangeable.[1]

In the second model, Q and Qbar are given mode **out** in the entity declaration, because they are not read inside the model. Q and Qbar are re-evaluated only if S or R or both are '0'. If both S and R are '1', the outputs retain their value.

Note that in the behavioural architecture, the **case** statement must, by definition, cover all possible values of S and R (hence the **when others** clause), but because we do not wish this branch to change the outputs, we include a **null** statement. The sequential **null** statement is equivalent to the concurrent **unaffected** clause or to omitting the final **else** in a **when** statement. Therefore it would have been possible to

---

[1]This restriction has been eased in VHDL2002 – see Appendix A.

declare Q and Qbar to have mode **out**, as in the second model, but to have the following architecture:

```
architecture dataflow of SR_latch2 is
begin
  Q <= '1' when R = '0' else
       '0' when S = '0' else
         unaffected;
  Qbar <= '1' when S = '0' else
          '0' when R = '0' else
            unaffected;
end architecture dataflow;
```

6.1.2 D latch

Because an SR latch can have both outputs at the same value, it is seldom if ever used. More useful is the D latch, as shown in Figure 6.2. The input of a D latch is transferred to the output if an enable signal is asserted. $1D$ indicates a dependency of the D input on control signal 1 ($C1$). The $\overline{Q}$ output is not shown.

A behavioural VHDL model of a D latch is

```
library IEEE;
use IEEE.std_logic_1164.all;

entity D_latch is
  port (D, Enable : in std_logic;
        Q : out std_logic);
end entity D_latch;

architecture behavioural of D_latch is
begin
p0: process (D, Enable) is
  begin
    if (Enable = '1') then
      Q <= D;
    end if;
  end process p0;
end architecture behavioural;
```

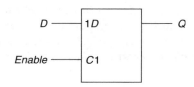

Figure 6.2 Level-sensitive D latch.

We could use one of a number of styles of modelling, but we will use the behavioural style for the following reasons:

- Declaring ports to have **inout** or **buffer** modes may make a dataflow style model difficult to use because such ports can be connected only to internal signals or other ports with the same mode. (As noted, this restriction on **buffer** ports has been eased in the 2002 standard of VHDL, but this new standard is unlikely to be universally adopted for some years.)

- We do not want to specify the structure of the latch – we are interested in modelling behaviour.

- The behavioural model is the only one that we can be certain will be synthesizable. The dataflow models of the SR latch, for example, are not supported by the 1076.6 RTL synthesis standard.

Notice that the sensitivity list of the process includes both the D signal and the Enable signal. The latch model is re-evaluated whenever either signal changes. If the Enable signal is high and the D input changes, that change will cause an immediate change in the Q output. This model is suitable for both simulation and synthesis; we will discuss why it would be synthesized to a level-sensitive D latch in Chapter 9.

6.2 Flip-flops

6.2.1 Edge-triggered D flip-flop

In the previous chapter the principle of synchronous sequential design was described. The main advantage of this approach to sequential design is that all changes of state occur at a clock *edge*. The clock edge is extremely short in comparison to the clock period and to propagation delays through combinational logic. In effect, a clock edge can be considered to be instantaneous. In VHDL, there are a number of ways to detect when a clock edge has occurred and hence to model an edge-triggered flip-flop.

The IEEE symbol for a positive edge-triggered D flip-flop is shown in Figure 6.3. Again, the number 1 shows the dependency of *D* on *C*. The triangle at the clock input denotes edge-sensitive behaviour. An inversion circle, or its absence, shows sensitivity to a negative or positive edge, respectively.

The simplest VHDL model of a positive edge-triggered D flip-flop is given below.

```
library IEEE;
use IEEE.std_logic_1164.all;
```

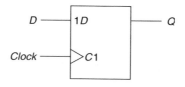

Figure 6.3   Positive edge-triggered D flip-flop.

```vhdl
entity D_FF is
  port (D, Clock : in std_logic;
        Q : out std_logic);
end entity D_FF;

architecture behavioural of D_FF is
begin
p0: process is
  begin
    wait until (Clock = '1');
    Q <= D;
  end process p0;
end architecture behavioural;
```

This form of the **process** statement was first introduced in the last chapter. The process does not have a sensitivity list; instead it contains a **wait** statement. A process may have either a sensitivity list or one or more **wait** statements, but not both. (A process could have neither a sensitivity list nor **wait** statements, in which case it will be continually re-evaluated. This is an error in VHDL.) The **wait** statement causes the process to suspend (i.e. to become inactive) until the condition specified in the **wait** statement becomes true. In this case the condition is that Clock *becomes* logical 1. Once the condition becomes true, the next statement is executed, the process restarts and then waits until the next rising clock edge. There are other forms of the **wait** statement that we will meet later.

An alternative model of a positive edge-triggered D flip-flop is as follows:

```vhdl
architecture alternative of D_FF is
begin
p0: process (Clock) is
  begin
    if (Clock = '1') then
      Q <= D;
    end if;
  end process p0;
end architecture alternative;
```

The process is re-evaluated whenever Clock changes state. The **if** statement is then used to check that Clock has changed to a logical 1. This model is not *exactly* equivalent to the previous model. The following model is, however, exactly equivalent to the second model:

```vhdl
architecture equivalent of D_FF is
begin
p0: process is
  begin
    if (Clock = '1') then
      Q <= D;
    end if;
```

```
    wait on Clock;
  end process p0;
end architecture equivalent;
```

A process with a sensitivity list is equivalent to a process with a **wait on** statement at the *end* of the process. The **wait on** form of a **wait** statement causes execution of the process to be suspended until there is a change in the signal (or signals) listed. It is important to understand that the first version of the flip-flop model differs from the second and third versions. At the beginning of a simulation, at time 0, each process in a VHDL description is executed as far as the first **wait** statement. Thus in the first model, the **wait** statement occurs at the beginning of the process, with the assignment statement following it. In the second and third versions, the assignment statement is before the **wait** statement, and thus would be executed at the start of a simulation if the **if** statement is true. In practice, the first and second models would both be interpreted by a synthesis tool as positive edge-triggered flip-flops. The third model would simulate as a positive edge-triggered flip-flop, but would probably not be correctly interpreted by a synthesis tool.

Similarly, a negative edge-triggered flip-flop can be modelled by detecting a transition to logical 0:

```
architecture neg_edge of D_FF is
begin
p0: process is
  begin
    wait until (Clock = '0');
    Q <= D;
  end process p0;
end architecture neg_edge;
```

6.2.2   Asynchronous set and reset

When power is first applied to a flip-flop its initial state is unpredictable. In many applications this is unacceptable, so flip-flops are provided with further inputs to set (or reset) their outputs to 1 or 0, as shown in Figure 6.4. Notice that the absence of any dependency on the clock implies asynchronous behaviour for *R* and *S*.

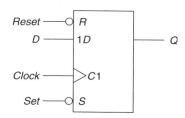

Figure 6.4   Positive edge-triggered D flip-flop with asynchronous reset and set.

These inputs should be used only to initialize a flip-flop. It is very bad practice to use these inputs to set the state of a flip-flop during normal system operation. The reason for this is that in synchronous systems, flip-flops change state only when clocked. The set and reset inputs are *asynchronous* and hence cannot be guaranteed to change an output at a particular time. This can lead to all sorts of timing problems. In general, keep all designs strictly synchronous or follow a structured asynchronous design methodology.

A VHDL model of a flip-flop with an asynchronous reset must respond to changes in the clock and in the reset input. Therefore we must use a process with a sensitivity list that includes both these signals. The **wait until** and **wait on** forms of the flip-flop model cannot be written to include checks on both inputs, so we use an **if** statement. The process will be activated whenever Clock or Reset changes. If we make the reset active low, Q is forced to 0 while the reset is 0, irrespective of the state of the clock. Therefore the first test in the **if** statement must be made on the Reset signal:

```
if (Reset = '0') then
  Q <= '0';
```

If this signal is not asserted, we then need to check whether we have a rising edge on the clock. Note, however, that at this point, simply checking to see whether the clock is at 1 is not sufficient, because the process might have been activated by the Reset signal changing to 1. Thus we have to check that the clock is at 1 and that it was a transition on the Clock signal that activated the process. This can be done by checking the 'EVENT attribute of the Clock signal. This attribute is true only if a change has occurred to that signal at the current time.

```
elsif (Clock = '1' and Clock'EVENT) then
  Q <= D;
```

Another way to detect an edge is to check the 'STABLE attribute. This is true if the signal has *not* changed at the current time.

```
elsif (Clock = '1' and not Clock'STABLE) then
  Q <= D;
```

The entire flip-flop model is now shown below.

```
library IEEE;
use IEEE.std_logic_1164.all;

entity D_FF_R is
  port (D, Clock, Reset : in std_logic;
        Q : out std_logic);
end entity D_FF_R;

architecture behavioural of D_FF_R is
begin
p0: process (Clock, Reset) is
```

```
    begin
      if (Reset = '0') then
        Q <= '0';
      elsif (Clock = '1' and Clock'EVENT) then
        Q <= D;
      end if;
    end process p0;
end architecture behavioural;
```

An asynchronous set can be described in a similar way (see Exercises).

It is possible for a flip-flop to have both an asynchronous set and reset. For example:

```
library IEEE;
use IEEE.std_logic_1164.all;

entity D_FF_RS is
  port (D, Clock, Reset, Set : in std_logic;
        Q : out std_logic);
end entity D_FF_RS;

architecture behavioural of D_FF_RS is
begin
p0: process (Clock, Reset, Set) is
  begin
    if (Set = '0') then
      Q <= '1';
    elsif (Reset = '0') then
      Q <= '0';
    elsif (Clock = '1' and Clock'EVENT) then
      Q <= D;
    end if;
  end process p0;
end architecture behavioural;
```

This may not correctly describe the behaviour of a flip-flop with asynchronous inputs because asserting both the asynchronous set and reset is usually considered an illegal operation. In this model, Q is forced to 1 if Set is 0, regardless of the Reset signal. Even if this model synthesizes correctly, we would still wish to check that this condition did not occur during a simulation. A technique to do this is described later in this chapter.

6.2.3   Rising_edge and falling_edge

All these flip-flop models detect a transition *to* a given state. If the clock were declared to be of type BIT, the only transitions of the clock that would be possible are 0 to 1 and 1 to 0. With the std_logic type, there are nine logic states and hence eight possible transitions to any given state. The flip-flop models shown will trigger on, for instance, an *H* to 1 transition. This is not how a real flip-flop behaves. We can specify a true 0 to 1 transition by checking both the current value of the clock and its last value. This is done using the 'LAST_VALUE attribute:

```vhdl
architecture true_edge of D_FF_R is
begin
p0: process (Clock, Reset) is
  begin
    if (Reset = '0') then
      Q <= '0';
    elsif (Clock = '1' and
           Clock'LAST_VALUE = '0' and Clock'EVENT) then
      Q <= D;
    end if;
  end process p0;
end architecture true_edge;
```

This has now defined a true 0 to 1 transition, but this form is not recognized by all synthesis tools. Further, an L to 1 (weak logic 0 to strong logic 1) transition would not be recognized in a simulation. The standard logic package simplifies all these cases by providing a rising_edge and a falling_edge function.

```vhdl
architecture r_edge of D_FF_R is
begin
p0: process (Clock, Reset) is
  begin
    if (Reset = '0') then
      Q <= '0';
    elsif rising_edge(Clock) then
      Q <= D;
    end if;
  end process p0;
end architecture r_edge;
```

It is strongly suggested that these functions be used to detect clock edges.

6.2.4 Synchronous set and reset and clock enable

Flip-flops may have synchronous set and reset functions as well as, or instead of, asynchronous set or reset inputs. A synchronous set or reset takes effect only at a clock edge. Thus a VHDL model of such a function must include a check on the set or reset input after the clock edge has been checked. It is not necessary to include synchronous set or reset inputs in the process sensitivity list because the process is activated only at a clock edge. This is shown in IEEE notation in Figure 6.5. R is now shown to be dependent on C and is therefore synchronous.

```vhdl
architecture synch_reset of D_FF_R is
begin
p0: process (Clock) is
```

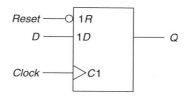

Figure 6.5 Positive edge-triggered D flip-flop with synchronous reset.

```
  begin
    if rising_edge(Clock) then
      if (Reset = '0') then
        Q <= '0';
      else
        Q <= D;
      end if;
    end if;
  end process p0;
end architecture synch_reset;
```

Similarly, a flip-flop with a clock enable signal may be modelled with that signal checked after the edge detection. In Figure 6.6, the dependency notation shows that *C* is dependent on *G*, and *D* is dependent on (the edge-triggered behaviour of) *C*.

```
library IEEE;
use IEEE.std_logic_1164.all;

entity D_FF_E is
    port (D, Clock, Enable : in std_logic;
          Q : out std_logic);
end entity D_FF_E;

architecture behavioural of D_FF_E is
begin
p0: process (Clock) is
  begin
    if rising_edge(Clock) then
      if (Enable = '1') then
```

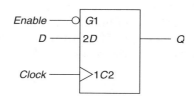

Figure 6.6 Positive edge-triggered flip-flop with clock enable.

```
        Q <= D;
      end if;
    end if;
  end process p0;
end architecture behavioural;
```

A synthesis system is likely to interpret this as a flip-flop with a clock enable. The following model is likely to be interpreted differently, although it appears to have the same functionality:

```
architecture gated_clock of D_FF_E is
  signal ce : std_logic;
begin
  ce <= Enable and Clock;
p0: process (ce) is
  begin
    if rising_edge(ce) then
      Q <= D;
    end if;
  end process p0;
end architecture gated_clock;
```

Again, the *D* input is latched if Enable is true and there is a clock edge. This time, however, the clock signal passes through an AND gate and hence is delayed. The *D* input is also latched if the clock is true and there is a rising edge on the Enable signal! This is another example of design that is not truly synchronous and that is therefore liable to timing problems. This style of design should generally be avoided, although for low-power applications the ability to turn off the clock inputs to flip-flops can be useful.

6.2.5   Timing and logic checks

In developing digital systems, we assume certain types of behaviour such as discrete logic levels. A further assumption, discussed in Chapter 12, is that only one input to a flip-flop can change at one time. For example, the *D* input to a flip-flop must have changed and be stable for a short period before the clock changes. Failure to observe this condition may result in an unpredictable output. In the worst case, the output of a flip-flop can exist in a *metastable* state somewhere between logical 1 and logical 0 for an *indeterminate* time. This unpredictability is not desirable. If we were verifying our designs by simulation, it would clearly be helpful if we were alerted to possible timing problems and to illegal combinations of inputs. VHDL provides the **assert** statement to generate warning messages. The **assert** statement is ignored by synthesis tools.

The form of an **assert** statement is as follows:

```
assert condition
  report message
  severity level;
```

The `condition` is a Boolean expression that we normally expect to be true. If the condition is `false` the message in the **report** part is printed. The severity level may be NOTE, WARNING, ERROR or FAILURE. An error or failure will usually cause the simulation to halt at that point. The **report** and/or the **severity** clause may be omitted. It is also possible to omit the **assert** part, in which case the message in the **report** part will always be printed. **Assert** statements may be included in sequential code or in concurrent code. The difference is that a concurrent **assert** will be activated only when one of the signals in the condition clause changes, while the sequential **assert** will be evaluated whenever it is reached in a process or other sequential block.

In Section 6.2.2 it was noted that an asynchronous **set** and **reset** should not both be at logical 0. This condition could be verified by the following **assert** statement:

```
assert (Set = '1' or Reset = '1')
  report "Set and Reset are both asserted"
  severity WARNING;
```

Thus if both inputs are at 0, the message is printed. Because we are stating what we expect to be true, the logic may appear to be counter-intuitive. We could equally state the condition that we are checking for and invert it:

```
assert (not(Set = '0' and Reset = '0'))
```

If we wish to check that the *D* input has stabilized before the clock input changes, we can use a form of the `'STABLE` attribute:

```
assert (not(Clk = '1' and Clk'EVENT and not D'STABLE(3 NS)))
  report "Setup time violation"
  severity WARNING;
```

Thus, we expect that the condition that there has been a clock edge *and D* has *not* been stable for 3 ns is *not* normally true.

The hold time of a flip-flop is defined as the time after a clock edge for which a data input must be stable. This can be similarly defined:

```
assert (not(Clk = '1' and D'EVENT and not Clk'STABLE(5 NS)))
  report "Hold time violation"
  severity WARNING;
```

The **assert** statement is *passive*, meaning that there is no signal assignment. Passive processes and statements may be included in the entity part of a declaration. The advantage of doing this is that the check applies to *all* architectures and does not have to be restated for every architecture. A model of a *D* flip-flop with an asynchronous reset and set, a clock enable, setup time and asynchronous input checks and propagation delays is shown below.

```
library IEEE;
use IEEE.std_logic_1164.all;
```

```
entity D_FF is
  generic (CQ_Delay, SQ_Delay, RQ_Delay: DELAY_LENGTH :=
           5 NS;
           Setup: DELAY_LENGTH := 3 NS);
  port (D, Clk, Set, Reset, Enable : in std_logic;
        Q : out std_logic);
begin
  assert (not(rising_edge(Clk) and not D'STABLE(Setup)))
    report "Setup time violation"
    severity WARNING;
end entity D_FF;

architecture behavioural of D_FF is
begin
p0: process (Clk, Set, Reset) is
  begin
    assert (not(Set = '0' and Reset = '0'))
      report "Set and Reset are both asserted"
      severity ERROR;
    if Set = '0' then
      Q <= '1' after SQ_Delay;
    elsif Reset = '0' then
      Q <= '0' after RQ_Delay;
    elsif rising_edge(Clk) then
      if (Enable = '1') then
        Q <= D after CQ_Delay;
      end if;
    end if;
  end process p0;
end architecture behavioural;
```

6.3 JK and T flip-flops

A D flip-flop registers its input at a clock edge, making that value available during the next clock cycle. JK and T flip-flops change their output states at the clock edge in response to their inputs and to their present states. Truth tables for D, JK and T flip-flops are shown below.

D	Q^+	$\overline{Q}^+$		J	K	Q^+	$\overline{Q}^+$		T	Q^+	$\overline{Q}^+$
0	0	1		0	0	Q	$\overline{Q}$		0	Q	$\overline{Q}$
1	1	0		0	1	0	1		1	$\overline{Q}$	Q
				1	0	1	0				
				1	1	$\overline{Q}$	Q				

Both the Q and $\overline{Q}$ outputs are shown. Symbols for D, JK and T flip-flops with both outputs and with a reset are shown in Figure 6.7.

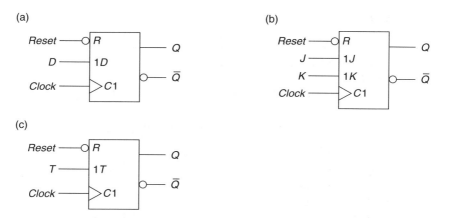

Figure 6.7   (a) D flip-flop; (b) JK flip-flop; (c) T flip-flop.

Before writing models for the JK and T flip-flops we will consider how a D flip-flop with both outputs would be modelled. The entity declaration for a D flip-flop with Q and $\overline{Q}$ outputs is:

```
library IEEE;
use IEEE.std_logic_1164.all;

entity D_FF is
  port (D, Clock, Reset : in std_logic;
        Q, Qbar : out std_logic);
end entity D_FF;
```

We cannot simply write `Qbar <= not Q;` to generate the `Qbar` output because Q is declared to be an output and cannot therefore be read. Q could be declared to have mode **buffer**, but that might restrict how the model could be used. Therefore we will store the state of the flip-flop internally. This can be done with a signal or with a variable.

```
architecture sig of D_FF is
  signal state : std_logic;
begin
p0: process (Clock, Reset) is
  begin
    if (Reset = '0') then
      state <= '0';
    elsif rising_edge(Clock) then
      state <= D;
    end if;
  end process p0;
  Q <= state;
  Qbar <= not state;
end architecture sig;
```

```
architecture var of D_FF is
begin
p0: process (Clock, Reset) is
    variable state : std_logic;
  begin
    if (Reset = '0') then
      state := '0';
    elsif rising_edge(Clock) then
      state := D;
    end if;
    Q <= state;
    Qbar <= not state;
  end process p0;
end architecture var;
```

In the first case, state is declared as a **signal**, outside the process. Two concurrent assignments to Q and Qbar are made outside (and concurrently with) the process. These two assignments are made outside the process because signals are updated in a process *after* that process suspends. Therefore, if we had the sequence

```
state <= D;
Q <= state;
```

inside a process, state would be updated with the new value of D at the same time as Q is updated with the last value of state. Hence two clock edges would be required to update the value of Q to a new value of D. A synthesis tool would interpret a sequence of two signal assignments such as this as implying the existence of *two* flip-flops in series.

In the second case, state is declared as a **variable** inside the process. Variables can be declared only inside a process or a subprogram. (Signals *cannot* be declared inside a process or subprogram.) The assignment to the variable, state, uses a different symbol (:=) from a signal assignment (<=). Unlike a signal assignment, a variable assignment takes effect immediately. Thus the new value of a variable is available in subsequent lines of a process. The assignments to Q and Qbar now take place inside the process and the new value of D is assigned to Q at a single clock edge. These assignments cannot be done outside, because the state variable is visible only inside the process.

Both the JK and T flip-flops can use a signal or variable to hold the internal state in the same way:

```
library IEEE;
use IEEE.std_logic_1164.all;

entity JK_FF is
  port (J, K, Clock, Reset : in std_logic;
        Q, Qbar : out std_logic);
end entity JK_FF;

architecture sig of JK_FF is
  signal state : std_logic;
```

```vhdl
begin
p0: process (Clock, Reset) is
  begin
    if (Reset = '0') then
      state <= '0';
    elsif rising_edge(Clock) then
      case std_logic_vector'(J, K) is
        when "11" =>
          state <= not state;
        when "10" =>
          state <= '1';
        when "01" =>
          state <= '0';
        when others =>
          null;
      end case;
    end if;
  end process p0;
  Q <= state;
  Qbar <= not state;
end architecture sig;

library IEEE;
use IEEE.std_logic_1164.all;

entity T_FF is
    port (T, Clock, Reset : in std_logic;
          Q, Qbar : out std_logic);
end entity T_FF;

architecture var of T_FF is
begin
p0: process (Clock, Reset) is
    variable state : std_logic;
  begin
    if (Reset = '0') then
      state := '0';
    elsif rising_edge(Clock) then
      if T = '1' then
        state := not state;
      end if;
    end if;
    Q <= state;
    Qbar <= not state;
  end process p0;
end architecture var;
```

A **case** statement determines the internal state of the JK flip-flop. The selector of the **case** statement is formed by aggregating the *J* and *K* inputs and qualifying the result as a std_logic_vector. The **when others** default clause covers the '00' case and other undefined values (a **case** statement must cover all possible values of the selector). The **null** statement does nothing, so the value of state is retained.

The internal state of the T flip-flop is held as a variable. Note that the value of a variable is retained between activations of a process.

6.4 Registers and shift registers

6.4.1 Multiple bit register

A D flip-flop is a one-bit register. Thus if we want a register with more than one bit, we simply need to define a set of D flip-flops using vectors:

```
library IEEE;
use IEEE.std_logic_1164.all;

entity reg is
  generic (n : NATURAL := 4);
  port (D : in std_logic_vector(n-1 downto 0);
        Clock, Reset : in std_logic;
        Q : out std_logic_vector(n-1 downto 0));
end entity reg;

architecture behavioural of reg is
begin
p0: process (Clock, Reset) is
  begin
    if (Reset = '0') then
      Q <= (others => '0');
    elsif rising_edge(Clock) then
      Q <= D;
    end if;
  end process p0;
end architecture behavioural;
```

The IEEE symbol for a 4-bit register is shown in Figure 6.8. Note that the common signals are contained in a control block.

6.4.2 Shift registers

An extension of the above model of a register includes the ability to shift the bits of the register to the left or to the right. For example, a sequence of bits can be converted into a word by shifting the bits into a register, and moving the bits along at each clock edge.

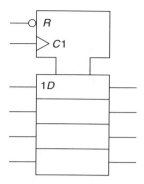

Figure 6.8 Four-bit register.

After a sufficient number of clock edges, the bits of the word are available as a single word. This is known as a *serial-in, parallel-out* (SIPO) register.

```
library IEEE;
use IEEE.std_logic_1164.all;

entity sipo is
  generic(n : NATURAL := 8);
  port(a : in std_logic;
       q : out std_logic_vector(n-1 downto 0);
       clk : in std_logic);
end entity sipo;

architecture rtl of sipo is
begin
p0: process (clk) is
    variable reg : std_logic_vector(n-1 downto 0);
  begin
    if rising_edge(clk) then
      reg := reg(n-2 downto 0) & a;
      q <= reg;
    end if;
  end process p0;
end architecture rtl;
```

At each clock edge, the bits of the register are moved along by one, and the input, a, is shifted into the 0th bit. The assignment to reg does this by assigning bits $n - 2$ to 0 to bits $n - 1$ to 1, respectively, and concatenating a to the end of the assignment. The old value for bit $n - 1$ is lost.

A more general shift register is the universal shift register. This can shift bits to the left or to the right, and can load an entire new word in parallel. To do this, two control bits are needed. The IEEE symbol is shown in Figure 6.9.

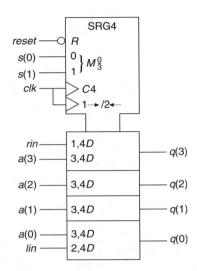

Figure 6.9   Universal shift register.

S_1S_0	Action
00	Hold
01	Shift right
10	Shift left
11	Parallel load

There are four control modes shown by $M\frac{0}{3}$. The clock signal is split into two for convenience. Control signal 4 is generated and in modes 1 and 2 a shift left or shift right operation, respectively, is performed. 1,4D means that a D-type operation occurs in mode 1 when control signal 4 is asserted.

```vhdl
library IEEE;
use IEEE.std_logic_1164.all;

entity usr is
  generic(n : NATURAL := 8);
  port(a : in std_logic_vector(n-1 downto 0);
       lin, rin : in std_logic;
       s : in std_logic_vector(1 downto 0);
       clk, reset : in std_logic;
       q : out std_logic_vector(n-1 downto 0));
end entity usr;

architecture rtl of usr is
begin
p0: process(clk, reset) is
    variable reg : std_logic_vector(n-1 downto 0);
```

```
    begin
      if (reset = '0') then
        reg := (others => '0');
      elsif rising_edge(clk) then
        case s is
          when "11" =>
            reg := a;
          when "10" =>
            reg := reg(n-2 downto 0) & lin;
          when "01" =>
            reg := rin & reg(n-1 downto 1);
          when others =>
            null;
        end case;
      end if;
      q <= reg;
    end process p0;
  end architecture rtl;
```

The shift operations are done by taking the lowest $(n - 1)$ bits and concatenating the leftmost input (shift left) or by taking the upper $(n - 1)$ bits concatenated to the rightmost input (shift right). It would be possible to use the shift operators, but in practice they are not needed.

6.5 Counters

Counters are used for a number of functions in digital design, e.g. counting the number of occurrences of an event, storing the address of the current instruction in a program, or generating test data. Although a counter typically starts at zero and increments monotonically to some larger value, it is also possible to use different sequences of values, which can result in simpler combinational logic.

6.5.1 Binary counter

A binary counter is a counter in the intuitive sense. It consists of a register of a number of D flip-flops, the content of which is the binary representation of a decimal number. At each clock edge the content of the counter is increased by one, as shown in Figure 6.10. We can easily model this in VHDL, using the numeric_std package to provide the '+' operator. The reset operation is shown in Figure 6.10 as setting the contents (CT) to 0. The weight of each stage is shown in brackets.

```
library IEEE;
use IEEE.std_logic_1164.all, IEEE.numeric_std.all;

entity counter is
  generic(n : NATURAL := 4);
```

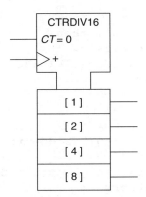

Figure 6.10   Binary counter.

```
   port(clk : in std_logic;
        reset : in std_logic;
        count : out std_logic_vector(n-1 downto 0));
end entity counter;

architecture rtl of counter is
begin
p0: process (clk, reset) is
     variable cnt : unsigned(n-1 downto 0);
  begin
    if reset = '1' then
      cnt := (others => '0');
    elsif rising_edge(clk) then
      cnt := cnt + 1;
    end if;
    count <= std_logic_vector(cnt);
  end process p0;
end architecture rtl;
```

Note that the contents of the counter are stored as a variable inside a process. The variable has type unsigned (allowing the + operator to be used). The contents of the counter are assigned to a signal and converted into a std_logic_vector. Note that the + operator does not generate a carry out. Thus when the counter has reached its maximum integer value (all 1s) the next clock edge will cause the counter to 'wrap round' and its next value will be zero (all 0s). We could modify the counter to generate a carry out, but in general counters are usually designed to detect the all-1s state and to output a signal when that state is reached. A carry out signal would be generated one clock cycle later. It is trivial to modify this counter to count down, or to count by a value other than one (possibly defined by a generic – see the Exercises at the end of this chapter).

The advantage of describing a counter in VHDL is that the underlying combinational next state logic is hidden. For a counter with eight or more bits, the combinational logic can be very complex, but a synthesis system will generate that logic automatically. A simpler form of binary counter is the ripple counter. An example of a ripple counter using T flip-flops is described in VHDL below, using the T flip-flop of Section 6.3. The entity description is the same as above.

```
architecture ripple of counter is
  signal carry : std_logic_vector(n downto 0);
begin
  carry(0) <= clk;
  g0 : for i in 0 to n-1 generate
        ti: entity WORK.T_FF port map('1', carry(i), reset,
                                      count(i), carry(i + 1));
    end generate g0;
end architecture ripple;
```

Note that the T input is held at a constant value in the description. When simulated using the T flip-flop model, above, this circuit behaves identically to the RTL model. A more realistic model of a T flip-flop would, however, contain propagation delays:

```
architecture delayed of T_FF is
begin
p0: process (Clock, Reset) is
    variable state : std_logic;
  begin
    if (Reset = '0') then
      state := '0';
    elsif rising_edge(Clock) then
      if T = '1' then
        state := not state;
      end if;
    end if;
    Q <= state after 5 NS;
    Qbar <= not state after 5 NS;
  end process p0;
end architecture delayed;
```

Now a simulation of the ripple counter reveals its asynchronous nature. The second flip-flop is clocked from the $\overline{Q}$ output of the first flip-flop, as shown in Figure 6.11. A change in this output is delayed by 5 ns relative to the clock. Hence, the second flip-flop is clocked by a signal 5 ns behind the true clock. With further counter stages, the delay is increased. Further, it can be seen that incorrect intermediate values are generated. Provided the clock speed is sufficiently slow, a ripple counter can be used instead of a synchronous counter, but in many applications a synchronous counter is preferred.

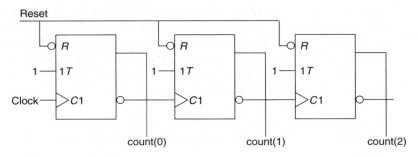

Figure 6.11   Ripple counter.

6.5.2   Johnson counter

A Johnson counter (also known as a Möbius counter – after a Möbius strip: a strip of paper formed into a circle with a single twist, resulting in a single surface) is built from a shift register with the least significant bit inverted and fed back to the most significant bit, as shown in Figure 6.12.

An n-bit binary counter has 2^n states. An n-bit Johnson counter has $2n$ states. The advantage of a Johnson counter is that it is simple to build (like a ripple counter), but is synchronous. The disadvantage is the large number of unused states that form an autonomous counter in their own right. In other words, we have the intended counter and a *parasitic* state machine coexisting in the same hardware. Normally, we should be unaware of the parasitic state machine, but if the system somehow entered one of the unused states, the subsequent behaviour might be unexpected. A VHDL description of a Johnson counter is shown below. The entity description is the same as for the binary counter.

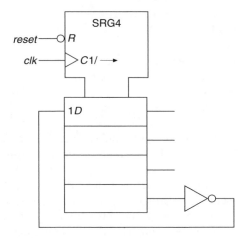

Figure 6.12   Johnson counter.

```vhdl
architecture johnson of counter is
begin
p0: process (clk, reset) is
    variable reg : std_logic_vector(n-1 downto 0);
  begin
    if reset = '1' then
      reg := (others => '0');
    elsif rising_edge(clk) then
      reg := not reg(0) & reg(n-1 downto 1);
    end if;
    count <= reg;
  end process p0;
end architecture johnson;
```

The counting sequence of a 4-bit counter, together with the sequence belonging to the parasitic state machine, is shown in the table below. Whatever the size of n, the unused states form a single parasitic counter with $2^n - 2n$ states.

Normal counting sequence	Parasitic counting sequence
0000	0010
1000	1001
1100	0100
1110	1010
1111	1101
0111	0110
0011	1011
0001	0101

Both sequences repeat but do not intersect at any point. The parasitic set of states of a Johnson counter should never occur, but if one of the states did occur somehow, perhaps because of a power supply glitch or because of some asynchronous input, the system can never return to its normal sequence. One solution to this is to make the counter *self-correcting*. It would be possible to detect every one of the parasitic states and to force a synchronous reset, but for an n-bit counter that is difficult. An easier solution is to note that the only legal state with a 0 in both the most significant and least significant bits is the all-zeros state. On the other hand, three of the parasitic states have zeros in those positions. Provided that we are happy to accept that if the system does enter an illegal state it does not have to correct itself immediately, but can re-enter the normal counting sequence after 'a few' clock cycles, we can simply detect any states that have a 0 at the most and least significant bits and force the next state to be '1000' or its n-bit equivalent.

```vhdl
architecture self_correcting_johnson of counter is
begin
p0: process (clk, reset) is
    variable reg : std_logic_vector(n-1 downto 0);
```

```
begin
  if reset = '1' then
    reg := (others => '0');
  elsif rising_edge(clk) then
    if reg(n-1) = '0' and reg(0) = '0' then
      reg := (others => '0');
      reg(n-1) := '1';
    else
      reg := not reg(0) & reg(n-1 downto 1);
    end if;
  end if;
  count <= reg;
end process p0;
end architecture self_correcting_johnson;
```

6.5.3 Linear feedback shift register

Another counter that is simple in terms of next-state logic is the linear feedback shift register (LFSR). This has $2^n - 1$ states in its normal counting sequence. The sequence of states appears to be random, hence the other name for the register: pseudo-random sequence generator (PRSG). The next-state logic is formed by exclusive OR gates as shown in Figure 6.13.

There are a large number of possible feedback connections for each value of n that give the maximal length ($2^n - 1$) sequence, but it can be shown that no more than four feedback connections (and hence three exclusive OR gates) are ever needed. The single state missing from the sequence is the all-0s state. Hence the asynchronous initialization should be a 'set'. As with the Johnson counter, the LFSR could be made self-correcting. A VHDL model of an LFSR valid for certain values of n is shown below.

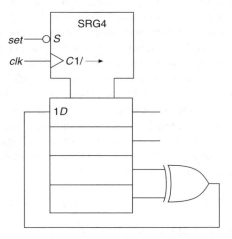

Figure 6.13 LFSR.

The main advantage of using an LFSR as a counter is that nearly the full range of possible states ($2^n - 1$) can be generated using simple next-state logic. Moreover, the pseudo-random sequence can be exploited for applications such as coding.

In the VHDL model, the feedback connections for LFSRs with 1 to 36 stages are defined in a table, taps. The table type is declared as a two-dimensional array. It is possible to define arrays with three or more dimensions in a similar manner. Note that the generic parameter, n, that defines the size of the LFSR is limited to the range 1 to 36 (with a default value of 8). Any attempt to use this model for a larger LFSR would result in a compilation error. The table has four entries per row for four feedback connections or fewer. If fewer than four feedback connections are needed, -1 is put in each of the spare places. (In all the cases listed, the last valid feedback connection in the corresponding table row is a 0. We could have been 'clever' and used this fact in our model. Hence the table would not have needed a special value of '-1' to indicate unused feedback connections. This would, however, be very poor software engineering practice. The algorithm breaks down if one piece of data does not fit the pattern and debugging code that relies on accidental patterns in data is very difficult.) Thus to construct the feedback connection for a particular size of LFSR, the stages of the LFSR referenced in the taps table are XORed together using a **loop**. Entries in the taps table with a value of -1 are ignored.

```vhdl
library IEEE;
use IEEE.std_logic_1164.all;

entity lfsr is
  generic(n : INTEGER range 1 to 36 := 8);
  port(reset : in std_logic;
       q : out std_logic_vector(n-1 downto 0);
       clk : in std_logic);
end entity lfsr;

architecture rtl of lfsr is
  type tap_table is array (1 to 36, 1 to 4) of
       INTEGER range -1 to 36;
  constant taps : tap_table :=(
                              ( 0, -1, -1, -1),   -- 1
                              ( 1,  0, -1, -1),   -- 2
                              ( 1,  0, -1, -1),   -- 3
                              ( 1,  0, -1, -1),   -- 4
                              ( 2,  0, -1, -1),   -- 5
                              ( 1,  0, -1, -1),   -- 6
                              ( 1,  0, -1, -1),   -- 7
                              ( 6,  5,  1,  0),   -- 8
                              ( 4,  0, -1, -1),   -- 9
                              ( 3,  0, -1, -1),   -- 10
                              ( 2,  0, -1, -1),   -- 11
                              ( 7,  4,  3,  0),   -- 12
                              ( 4,  3,  1,  0),   -- 13
```

```
                                        ( 12, 11,   1,   0),  -- 14
                                        (  1,  0,  -1,  -1),  -- 15
                                        (  5,  3,   2,   0),  -- 16
                                        (  3,  0,  -1,  -1),  -- 17
                                        (  7,  0,  -1,  -1),  -- 18
                                        (  6,  5,   1,   0),  -- 19
                                        (  3,  0,  -1,  -1),  -- 20
                                        (  2,  0,  -1,  -1),  -- 21
                                        (  1,  0,  -1,  -1),  -- 22
                                        (  5,  0,  -1,  -1),  -- 23
                                        (  4,  3,   1,   0),  -- 24
                                        (  3,  0,  -1,  -1),  -- 25
                                        (  8,  7,   1,   0),  -- 26
                                        (  8,  7,   1,   0),  -- 27
                                        (  3,  0,  -1,  -1),  -- 28
                                        (  2,  0,  -1,  -1),  -- 29
                                        ( 16, 15,   1,   0),  -- 30
                                        (  3,  0,  -1,  -1),  -- 31
                                        ( 28, 27,   1,   0),  -- 32
                                        ( 13,  0,  -1,  -1),  -- 33
                                        ( 15, 14,   1,   0),  -- 34
                                        (  2,  0,  -1,  -1),  -- 35
                                        ( 11,  0,  -1,  -1)); -- 36
  begin
  p0: process (clk, reset) is
       variable reg : std_logic_vector(n-1 downto 0);
       variable feedback : std_logic;
     begin
       if reset = '1' then
         reg := (others => '1');
       elsif rising_edge(clk) then
         feedback := reg(taps(n, 1));
         for i in 2 to 4 loop
           if taps(n, i) >= 0 then
             feedback := feedback xor reg(taps(n, i));
           end if;
         end loop;
         reg := feedback & reg(n-1 downto 1);
       end if;
       q <= reg;
     end process p0;
  end architecture rtl;
```

6.6 Memory

Computer memory is often classified as ROM (read-only memory) and RAM (random access memory). These are to some extent misnomers – ROM is random access and RAM is better thought of as read and write memory. RAM can further be divided into SRAM (static RAM) and DRAM (dynamic RAM). Static RAM retains its contents while power is applied to the system. Dynamic RAM uses capacitors to store bits, which means that the capacitance charge can leak away with time. Hence DRAM needs refreshing intermittently.

6.6.1 ROM

The contents of a ROM chip are defined once. Hence we can use a constant array to model a ROM device in VHDL. Below is the seven-segment decoder from Chapter 4 described as a ROM.

```
library IEEE;
use IEEE.std_logic_1164.all;

entity rom16x7 is
   port (address : in INTEGER range 0 to 15;
         data : out std_logic_vector (6 downto 0));
end entity rom16x7;

architecture sevenseg of rom16x7 is
   type rom_array is array (0 to 15) of
     std_logic_vector (6 downto 0);
   constant rom : rom_array := ("1110111",
                                "0010010",
                                "1011101",
                                "1011011",
                                "0111010",
                                "1101011",
                                "1101111",
                                "1010010",
                                "1111111",
                                "1111011",
                                "1101101",
                                "1101101",
                                "1101101",
                                "1101101",
                                "1101101",
                                "1101101");
begin
   data <= rom(address);
end architecture sevenseg;
```

The value contained in the ROM for a given address is available at the data output after the address has been set up – if we are interested in the timing characteristics, this delay could be modelled in the VHDL code:

```
data <= rom(address) after 5 NS;
```

Because no values can be written into the ROM, we can think of the device as combinational logic. In general, combinational logic functions can be implemented directly in ROM. Programmable forms of ROM are available (EPROM – electrically programmable ROM), but such devices require the application of a large negative voltage (-12 V) to a particular pin of the device. Such functionality is not modelled, as it does not form part of the normal operating conditions of the device.

6.6.2   Static RAM

A static RAM may be modelled in much the same way as a ROM. Because data may be stored in the RAM as well as read from it, the data signal is declared to be of mode **inout**. In addition, three control signals are provided. The first, CS (Chip Select), is a general control signal to enable a particular RAM chip. The address range, in this example, is 0 to 15. If we were to use, say, four identical chips to provide RAM with an address range of 0 to 63 (6 bits), the upper two address bits would be decoded such that at any one time exactly one of the RAM chips is enabled by its CS signal. Hence if the CS signal is not enabled, the data output of the RAM chip should be in the high-impedance state. The other two signals are OE (Output Enable) and WE (Write Enable). Only one of these two signals should be asserted at one time. Data is either read from the RAM chip when the OE signal is asserted, or written to the chip if the WE signal is asserted. If neither signal is asserted, the output remains in the high-impedance state. All the control signals are active low.

Like the ROM, the memory array is modelled as an array, this time as a variable in a process. The default initial value of a signal or variable is the leftmost value of its type. In this case the leftmost value of `std_logic` is 'U', so the entire memory array is initialized to 'U'. Again no timing has been included, but this could be easily done.

```
library IEEE;
use IEEE.std_logic_1164.all;

entity RAM16x8 is
  port (Address: in INTEGER range 0 to 15;
        Data: inout std_logic_vector(7 downto 0);
        CS, WE, OE: in std_logic);
end entity RAM16x8;

architecture RTL of RAM16x8 is
begin
p0: process (Address, CS, WE, OE, Data) is
    type ram_array is array (0 to 15) of
        std_logic_vector(7 downto 0);
    variable mem: ram_array;
```

```vhdl
  begin
    Data <= (others => 'Z');
    if CS = '0' then
      if OE = '0' then
        Data <= mem(Address);
      elsif WE = '0' then
        mem(Address) := Data;
      end if;
    end if;
  end process p0;
end architecture RTL;
```

6.6.3   Dynamic RAM

Dynamic RAM is fundamentally different from SRAM. In SRAM, memory cells are built from standard (CMOS) latches. In DRAM, bits are stored as charge on the capacitance of transistor gates. DRAM is therefore much more compact than SRAM. It is, on the other hand, slower. Therefore, if you look inside a PC, you will find SRAM used for the cache because it is fast and DRAM used for the main memory because it is compact and cheap. You will also notice that the SIMM arrays inside a PC commonly have nine chips. Each DRAM device is an array of, say, 16 M*bits*. An eight-bit data word is therefore stored across eight DRAM chips, with its parity bit in the ninth chip.

Because the data in a DRAM chip is stored as charge on a capacitor, the data needs to be refreshed every so often as the capacitor slowly discharges. Additional circuitry is required to periodically read the entire contents of the DRAM and to write it back. The DRAM model, below, does not model the refresh circuitry, but it does model another important feature of many DRAM chips. The address of a data bit is written in two pieces: its row, followed by its column in the memory array. The timing of signals to a DRAM chip is therefore critical; a simplified timing diagram is shown in Figure 6.14.

Note that the DRAM model should be used only for simulation. It is not possible to fabricate DRAM devices on standard CMOS technology, and in any case a synthesis system would attempt to implement this model as SRAM!

```vhdl
library IEEE;
use IEEE.std_logic_1164.all;

entity DRAM8MBit is
  port(Address: in INTEGER range 0 to 2**10 - 1;
       Data: inout std_logic_vector(7 downto 0);
       RAS,CAS,WE: in std_logic);
end entity DRAM8MBit;

architecture behaviour of DRAM8Mbit is
begin
p0: process (RAS, CAS, WE) is
```

```
        type dram_array is array (0 to 2**20 - 1) of
            std_logic_vector(7 downto 0);
        variable row_address: INTEGER range 0 to 2**10 - 1;
        variable mem_address: INTEGER range 0 to 2**20 - 1;
        variable mem: dram_array;
    begin
        Data <= (others => 'Z');
        if falling_edge(RAS) then
          row_address := Address;
        elsif falling_edge(CAS) then
          mem_address := row_address*2**10 + Address;
          if RAS = '0' and WE = '0' then
            mem(mem_address) := Data;
          end if;
        elsif CAS = '0' and RAS = '0' and WE = '1' then
          Data <= mem(mem_address);
        end if;
    end process p0;
end architecture behaviour;
```

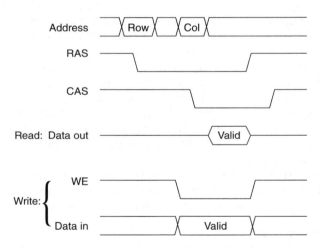

Figure 6.14 DRAM timing.

6.6.4   Synchronous RAM

Both the static and dynamic RAM models are asynchronous and intended for model-ling separate memory chips. Sometimes we wish to allocate part of an FPGA as RAM. In order for this to be synthesized correctly and for ease of use, it is best to make this RAM synchronous. Depending on the technology, there may be a variety of possible RAM structures, e.g. synchronous read, dual-port. Here, we will simply show how a basic synchronous RAM is modelled. This parameterizable example can be synthesized in most programmable technologies.

```vhdl
library IEEE;
use IEEE.std_logic_1164.all;
use IEEE.numeric_std.all;

entity SyncRAM is
  generic(M: NATURAL := 4;
          N: NATURAL := 8);
  port (Address: in std_logic_vector(M-1 downto 0);
        Data: in std_logic_vector(N-1 downto 0);
        WE, Clk: in std_logic;
        Qout: out std_logic_vector(N-1 downto 0));
end entity SyncRAM;

architecture RTL of SyncRAM is
  type ram_array is array (0 to 2**M-1) of
    std_logic_vector(N-1 downto 0);
  signal mem: ram_array;
begin
p0: process (Clk) is
  begin
    if rising_edge(clk) then
      if WE = '0' then
        mem(to_integer(unsigned(Address))) <= Data;
      end if;
    end if;
  end process p0;
  Qout <= mem(to_integer(unsigned(Address)));
end architecture RTL;
```

The structure of this code is almost identical to that of a flip-flop with an enable – in this case, the enable signal is the WE input. As with the SRAM example above, the Address input is converted into an unsigned integer to reference an array. This example can be extended to include an output enable and chip select, as above.

6.7 Sequential multiplier

Let us consider a multiplier for two's complement binary numbers. Multiplication, whether decimal or binary, can be broken down into a sequence of additions. A VHDL statement such as

```vhdl
q <= a * b;
```

where a and b are n-bit representations of (positive) integers, would be interpreted by a VHDL synthesis tool as a combinational multiplication requiring n^2 full adders.

If a and b are two's complement numbers, there also needs to be a sign adjustment. A combinational multiplier would take up a significant percentage of an FPGA for $n = 8$ and would require many FPGAs for $n = 16$.

The classic trade-off in digital design is between area and speed. In this case, we can significantly reduce the area required for a multiplier if the multiplication is performed over several clock cycles. Between additions, one of the operands of a multiplication operation has to be shifted. Therefore a multiplier can be implemented as a single n-bit adder and a shift register.

Two's complement numbers present a particular difficulty. It would be possible, but undesirable, to recode the operands as unsigned numbers with a sign bit. Booth's algorithm tackles the problem by treating an operand as a set of sequences of all 1s and all 0s. For example, -30 is represented as 100010. This is equal to $-2^5 + 2^2 - 2^1$. In other words, as each bit is examined in turn, from left to right, only a change from a 1 to a 0, or from a 0 to a 1, is significant. Hence, in multiplying b by a, each pair of bits of a is examined, so that if $a_i = 0$ and $a_{i-1} = 1$, b shifted by i places is added to the partial product. If $a_i = 1$ and $a_{i-1} = 0$, b shifted by i places is subtracted from the partial product. Otherwise no operation is performed. The VHDL model below implements this algorithm, but note that instead of shifting the operand to the left, the partial product is shifted to the right at each clock edge. A `ready` flag is asserted when the multiplication is complete.

```vhdl
library IEEE;
use IEEE.std_logic_1164.all, IEEE.numeric_std.all;

entity booth is
    generic(al : NATURAL := 16;
            bl : NATURAL := 16;
            ql : NATURAL := 32);
    port(ain : in std_logic_vector(al-1 downto 0);
         bin : in std_logic_vector(bl-1 downto 0);
         qout : out std_logic_vector(ql-1 downto 0);
         clk : in std_logic;
         load : in std_logic;
         ready : out std_logic);
end entity booth;

architecture rtl of booth is
begin
  process (clk) is
    variable count : INTEGER range 0 to al;
    variable pa : signed((al + bl) downto 0);
    variable a_1 : std_logic;
    alias p : signed(bl downto 0) is
      pa((al + bl) downto al);
  begin
    if (rising_edge(clk)) then
      if load = '1' then
```

```
            p := (others => '0');
            pa(al-1 downto 0) := signed(ain);
            a_1 := '0';
            count := al;
            ready <= '0';
          elsif count > 0 then
            case std_logic_vector'(pa(0), a_1) is
              when "01" =>
                p := p + signed(bin);
              when "10" =>
                p := p - signed(bin);
              when others => null;
            end case;
            a_1 := pa(0);
            pa := shift_right(pa, 1);
            count := count - 1;
          end if;
          if count = 0 then
            ready <= '1';
          end if;
          qout <= std_logic_vector(pa(al + bl - 1 downto 0));
        end if;
      end process;
end architecture rtl;
```

6.7.1   Aliases

It is sometimes convenient to give an alternative name to an object. In the example above, using the default vector length, pa is a 32-bit vector that contains the multiplier and the partial product. After each bit of the multiplier is used, it is shifted out of pa. At the same time the partial product grows. The 16 most significant bits of the partial product are used in the additions and subtractions. We could have written, for example,

```
pa((al+bl) downto al) := pa((al+bl) downto al) +
                         signed(bin);
```

It is more convenient to give an **alias** to the top 17 bits of pa:

```
alias p : signed(bl downto 0) is pa((al + bl) downto al);
```

Now p can be used directly as shown without having to specify a range each time it is used.

Similarly, a microprocessor might have 32-bit instructions that can be thought of as 8-bit opcodes and 24-bit operands. Again, the use of **alias**es can help to make the VHDL code clearer:

```
signal instruction : std_logic_vector (31 downto 0);
alias opcode : std_logic_vector (7 downto 0) is
                       instruction (31 downto 24);
alias operand : std_logic_vector (23 downto 0) is
                       instruction (23 downto 0);
```

6.8 Testbenches for sequential building blocks

In the last chapter, we considered testbenches for state machines. The function of those testbenches was to generate clock and reset signals, together with other data inputs. We considered how signals could be synchronized with the clock and how to monitor the states of output signals. In this chapter, we have seen how warnings can be generated when timing constraints of sequential elements are violated. While the testbench structures of the last chapter can be used to verify sequential building blocks in general, it would clearly also be desirable to simulate conditions under which timing violations can be checked. In this section, therefore, we will see how inputs can be generated at random times and how uncertainty can be added to the times of clock and data input changes. Later we will also use **assert** statements in testbenches to monitor outputs. First, however, we will see how to generate a clock with unequal high and low periods (mark to space ratio).

6.8.1 Asymmetric clock

This clock generation example in the last chapter models a clock with equal high and low periods. The following example shows a clock generator in which the frequency and mark/space ratio are parameters. Notice that the clock frequency is specified as a real number. This must be done for the example given to simulate correctly. If the frequency were specified as an integer, a mark period of, say, 45% would cause a clock to be generated with a period of 9 ns, and mark and space times of 4 ns and 5 ns, respectively, because of rounding errors. (Of course an integer could be converted to a real to achieve the same effect.) Note also that to get 4.5 and 5.5 ns, the resolution of the simulator must be set to 100 ps or less.

```
library IEEE;
use IEEE.std_logic_1164.all;

entity clock_gen is
  generic (Freq : REAL := 1.0E8; -- 100 MHz
           Mark: POSITIVE := 45); -- Mark length %
end entity clock_gen;

architecture cg of clock_gen is
  -- Mark time in ns
  constant ClockHigh :TIME := (REAL(Mark)*1.0E7/Freq)*NS;
  -- Space time in ns
  constant ClockLow :TIME := (REAL(100-Mark)*1.0E7/Freq)*NS;
  signal clock : std_logic := '0';
```

```vhdl
begin
  process is
  begin
    wait for ClockLow;
    clock <= '1';
    wait for ClockHigh;
    clock <= '0';
  end process;
end architecture cg;
```

6.8.2 Random pulse generator

The VHDL Math Package (1076.2) contains a pseudo-random number generator, uniform, in addition to many other functions. A pseudo-random number generator uses techniques similar to that of the LFSR described above. From a given pair of starting values, or seeds, the same sequence will always be generated. The function uniform returns a pseudo-random real number in the range 0.0 to 1.0.

We can use this function to generate, for example, a clock that changes on average every 10 ns, but with a random variation of ± 1 ns. As with the asymmetric clock, the resolution of the simulator must be set to 100 ps or better, otherwise the variation will always round to 0. This clock generator could be written as a process, as in previous examples. Having written such a generator, however, we might want to put it into a package for use elsewhere. We cannot put a process into a package, but we can write a procedure instead and put that into a package. Moreover a procedure can take parameters. For example, a complete package definition could be:

```vhdl
library IEEE;
use IEEE.std_logic_1164.all;
use IEEE.math_real.all;

package clocks is
  procedure noisy_clk (signal clock : out std_logic;
                       delay : in DELAY_LENGTH);
end package clocks;

package body clocks is
  procedure noisy_clk (signal clock : out std_logic;
                       delay : in DELAY_LENGTH) is
    variable seed1, seed2 : INTEGER := 42;
    variable rnd : REAL;
  begin
    loop
      clock <= '0';
      uniform (seed1, seed2, rnd);
      wait for delay + (rnd - 0.5) * NS;
      clock <= '1';
```

```
       uniform (seed1, seed2, rnd);
       wait for delay + (rnd - 0.5) * NS;
    end loop;
  end procedure noisy_clk;
end package body clocks;
```

Notice that there is a loop in the procedure. A procedure can be called concurrently or from within a process or another sub-program. If this procedure is called concurrently, it will execute once and finish. This is because a concurrent procedure call is equivalent to putting the procedure call inside a process, with **wait** statements triggered by the procedure input signals. As there are no input signals, the procedure would execute once. A suitable concurrent procedure call is:

```
c0: noisy_clk (clock, 7 NS);
```

If we want to model the behaviour of a clock and data whose relative timing is imperfect, it makes no difference if we apply the 'jitter' to the clock or to the data. We may also wish to model totally unsynchronized data. One widely used approximation to the random events arriving with a mean interval is the negative exponential function. We can use a random event generator as follows:

```
library IEEE;
use IEEE.std_logic_1164.all;
use IEEE.math_real.all;

entity testrnd is end entity testrnd;
architecture testrnd of testrnd is
  signal r : std_logic := '0';
  function negexp(rnd: REAL; t : TIME) return TIME is
  begin
    return INTEGER (-log(rnd)*(REAL(t / NS))) * NS;
  end function negexp;
begin
rand_wav: process is
    variable seed1, seed2 : INTEGER := 199;
    variable rnd : REAL;
  begin
    uniform (seed1, seed2, rnd);
    wait for negexp(rnd, 10 NS);
    r <= not r;
  end process rand_wav;
end architecture testrnd;
```

This will generate a signal that toggles with a mean interval of 10 ns, but with a variation between 0.0 and infinity. This contains another example of a sub-program definition, in this case a function. Both procedures and functions can be included in packages or in architecture declarations, as shown.

6.8.3   Checking responses with `assert` statements

In Section 6.2.5, the **assert** statement was introduced. In the examples given, the behaviour of a model was checked *within* the model. It is just as valid to use **assert** statements to check responses within a testbench. Unlike most other VHDL statements, an **assert** statement can be used concurrently (outside a process) or sequentially (within a process).

Suppose two versions of a design are to be compared (for example, a gate level model and an RTL model). The two versions take the same inputs. To avoid contention, their outputs must have different signals. The two versions might be instantiated and their outputs compared as follows:

```
v0 : entity WORK.design(struct) port map (in_a, in_b,
                                           out_s);
v1 : entity WORK.design(rtl) port map (in_a, in_b, out_b);
assert out_s = out_b
    report "Mismatch in behavioural and structural outputs"
    severity NOTE;
```

Although simple to implement, this approach is flawed because any timing differences, however slight, will generate warning messages. In practice, it is very likely that there will be some differences between the outputs of two models at different levels of abstraction, but these differences will probably not be significant.

Therefore, it is preferable to compare responses only at specified *strobe* times. For example, we might wish to check responses 5 ns after a rising clock edge. This could be done as follows. Again, we will put the check inside a procedure so that it can be reused later.

```
procedure check(signal clock, out_s, out_b : in
                std_logic) is
begin
  wait until clock = '1';
  wait for 5 NS;
  assert out_s = out_b
    report "Mismatch in behavioural and structural outputs"
    severity NOTE;
end procedure check;
```

As there are three inputs, note that we do not need a loop in this example – the procedure restarts whenever an input changes. All the inputs must be declared as signals. It is possible to pass the current value of a signal to a procedure by omitting the signal object declaration, in which case the value is passed as a constant. If this is done, the procedure will not restart when that signal changes.

Summary

In this chapter we have discussed a number of common sequential building blocks. VHDL models of these blocks have been written using processes. Most of these models are synthesizable using RTL synthesis tools. We have also considered further examples of testbenches for sequential circuits.

Further reading

As with combinational blocks, manufacturers' data sheets are a good source of information about typical SSI devices. In particular, it is worth looking in some detail at the timing specifications for SRAM and DRAM devices. The multiplier is an example of how relatively complicated computer arithmetic can be performed. Hennessy and Patterson have a good description of computer arithmetic units.

Exercises

6.1 Explain how positive edge-triggered behaviour can be described in VHDL, where the `std_logic` type is used to represent bits and only a '0' to '1' transition is considered valid.

6.2 Write a behavioural VHDL model of a negative edge-triggered D flip-flop with set and clear.

6.3 Include tests for setup and hold time violation in the D flip-flop of Exercise 6.2.

6.4 Write a VHDL model of a negative edge-triggered T-type flip-flop.

6.5 Write a VHDL model of a 10-state synchronous counter that asserts an output when the count reaches 10.

6.6 Write a VHDL model of an N-bit counter with a control input 'Up'. When the control input is '1' the counter counts up; when it is '0' the counter counts down. The counter should not, however, wrap round. When the all 1s or all 0s states are reached the counter should stop.

6.7 Write a VHDL model of an n-bit parallel to serial converter.

6.8 Write a VHDL testbench for this parallel to serial converter.

6.9 What are the advantages and disadvantages of ripple counters as opposed to synchronous counters?

6.10 Design a synchronous Johnson counter that visits eight distinct states in sequence. How would this counter be modified such that any unused states lead eventually to the normal counting sequence?

6.11 Design an LFSR which cycles through the following states: 001, 010, 101, 011, 111, 110, 100. Verify your design by simulation.

6.12 Explain the function of the device shown in Figure 6.15. Your answer should include a description of all the symbols.

6.13 Show, with a full circuit diagram, how the device of Figure 6.15 could be used to build a synchronous counter with 12 states. Show how a synchronous reset can be included.

6.14 What is the difference between static and dynamic RAM and what are the relative advantages and disadvantages of each type?

6.15 A 64k × 1 bit dynamic RAM has the following pins:
- $A_0 \ldots A_7$ (Address lines)
- D (Data Line)

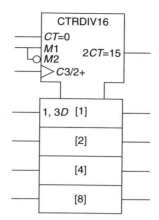

Figure 6.15   Device for Exercises 6.12 and 6.13.

- /RAS (Row Address Select)
- /CAS (Column Address Select)
- /WE (Write Enable)

Explain, with reference to these pins, how such RAM could be used in a micro-processor system requiring 256k × 8 bit memory. What timing considerations must be observed?

Chapter 7

Complex sequential systems

In the previous three chapters we have looked at combinational and sequential building blocks and at the design of state machines. The purpose of this chapter is to see how these various parts can be combined to build complex digital systems.

7.1 Linked state machines

In principle, any synchronous sequential system could be described by an ASM chart. In practice, this does not make sense. The states of a system, such as a microprocessor, include all the possible values of all the data that might be stored in the system. Therefore it is usual to partition a design in some way. In this chapter, we will show first how an ASM chart, and hence the VHDL model of the state machine, can be partitioned, and second how a conceptual split may be made between the *datapath* of a system, i.e. the components that store and manipulate data, and the state machine that controls the functioning of those datapath components.

A large body of theory covers the optimal partitioning of state machines. In practice, it is usually sufficient to identify components that can easily be separated from the main design and implemented independently. For example, let us consider again the traffic signal controller.

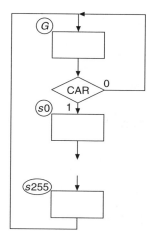

Figure 7.1 ASM chart of traffic signal controller including the timer.

If a car approaches the traffic signals on the minor road, a sensor is activated that causes the major road to have a red light and the minor road to have a green light for a fixed interval. Once that interval has passed, the major road has a green light again and the minor road has a red light. In Chapter 5, we simply assumed that a signal would be generated after the given interval had elapsed. Let us now assume that the clock frequency is such that the timed interval is completed in 256 clock cycles. We can draw an ASM chart for the entire system as shown in Figure 7.1 (states 1 to 254 and the outputs are not shown, for clarity). Although this is a simple example, the functionality of the system is somewhat lost in the profusion of states that implement a simple counting function. It would be clearer to separate the traffic light controller function from the timer.

One way of doing this is shown in Figure 7.2, in which there are two ASM charts. The ASM chart on the left is the traffic light controller, in which a signal, START, is asserted as a conditional output when a car is detected. This signal acts as an input to the second state machine, allowing that state machine to move from the IDLE state into the counting sequence. When the second state machine completes the counting sequence, the signal TIMED is asserted, which acts as an input to the first state machine, allowing the latter to move from state R to state G. The second state machine moves back into the IDLE state.

A state machine of the form of the right-hand state machine of Figure 7.2 can be thought of as a 'hardware subroutine'. In other words, any state machine may be partitioned in this way. Unlike a software subroutine, however, a piece of hardware must exist and must be doing something, even when it is not being used. Hence, the IDLE state must be included to account for the time when the 'subroutine' is not doing a useful task.

An alternative way of implementing a subsidiary state machine is shown in Figure 7.3. This version does not correspond to the 'hardware subroutine' model, but represents a conventional counter. The use of standard components will be discussed further in the next section.

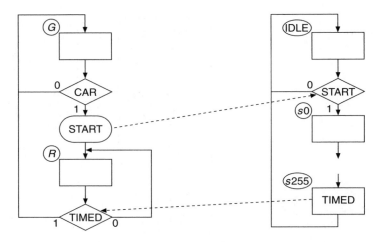

Figure 7.2 Linked ASM charts for traffic signal controller.

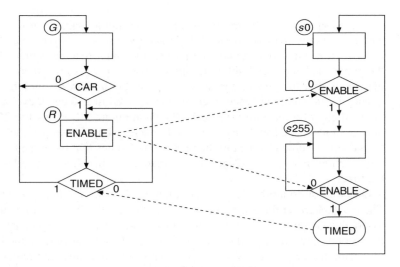

Figure 7.3 ASM chart of traffic signal controller with counter.

From the ASM chart of Figure 7.1 it is quite clear that the system takes 256 clock cycles to return to state G after a car has been detected. The sequence of operations may be harder to follow in Figure 7.3. In state $s255$, TIMED is asserted as a conditional output. This causes the left-hand state machine to move from state R to state G. In state R, ENABLE is asserted which allows the right-hand state machine to advance through its counting sequence. A timing diagram of this is shown in Figure 7.4.

At first glance this timing diagram may appear confusing. The ENABLE signal causes the TIMED signal to be asserted during the final state of the right-hand diagram. The TIMED signal causes the left-hand state machine to move from state R to state G.

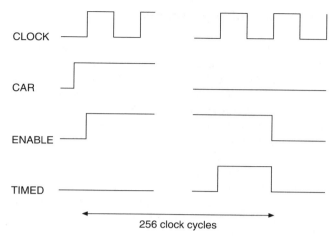

Figure 7.4   Timing diagram of linked ASM charts.

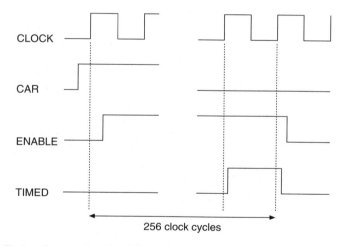

Figure 7.5   Timing diagram showing delays.

According to ASM chart convention, both these signals are asserted at the beginning of a state and deasserted at the end of a state. In fact, of course, the signals are asserted some time after a clock edge and also deasserted after a clock edge. Therefore, a more realistic timing diagram is given in Figure 7.5. The changes to TIMED and ENABLE happen after the clock edges. This, of course, is necessary in order to satisfy the setup and hold times of the flip-flops in the system. The clock speed is limited by the propagation delays through the combinational logic of both state machines. In that sense, a system divided into two or more state machines behaves no differently from a system implemented as a single state machine.

7.2 Datapath/controller partitioning

Although any synchronous sequential system can be designed in terms of one or more state machines, in practice this is likely to result in the 'reinvention of the wheel' on many occasions. For example, the right-hand state machine of Figure 7.3 is simply an 8-bit counter. Given this, it is obviously more effective to reuse an existing counter, either as a piece of hardware or as a VHDL model. It is therefore convenient to think of a sequential system in terms of the *datapath*, i.e. those components that have been previously designed (or that can be easily adapted) and that can be reused, and the *controller*, which is a design-specific state machine. A model of a system partitioned in this way is shown in Figure 7.6.

Returning to the example of Figure 7.3, it can be seen that the left-hand state machine corresponds to a controller, while the right-hand state machine, the counter, corresponds to the datapath. The TIMED signal is a status signal, as shown in Figure 7.6, while the ENABLE signal is a control signal. We will look at a more significant example of datapath/controller partitioning in Section 7.4.

The datapath would normally contain registers. As the functionality of the system is mainly contained in the datapath, the system can be described in terms of *register transfer operations*. These register transfer operations can be described using an extension of ASM chart notation. In the simplest case a registered output can be indicated as shown in Figure 7.7(a). This notation means that Z takes the value 1 *at the end* of the state indicated, and *holds that value* until it is reset. If, in this example, Z is reset to 0, and it is set to 1 only in the state shown, the registered output would be implemented as a flip-flop and multiplexer, as shown in Figure 7.7(b), or simply as an enabled flip-flop as shown in Figure 7.7(c). In either implementation, the ENABLE signal is asserted only when the ASM is in the indicated state. Thus the ASM chart could equally include the ENABLE signal, as shown in Figure 7.7(d).

A more complex example is shown in Figure 7.8. In state 00, three registers, B_0, B_1 and B_2, are loaded with inputs X_0, X_1 and X_2, respectively. Input A then determines whether a shift left, or multiply by 2, is performed ($A = 0$) or a shift right, or divide by 2 ($A = 1$) in the next state. If a divide by 2 is performed, the value of the least

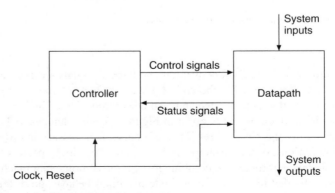

Figure 7.6 Controller/datapath partitioning.

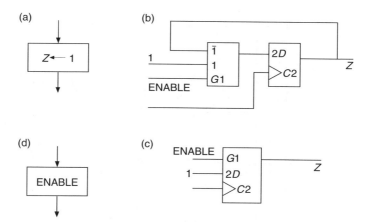

Figure 7.7   Extended ASM chart notation.

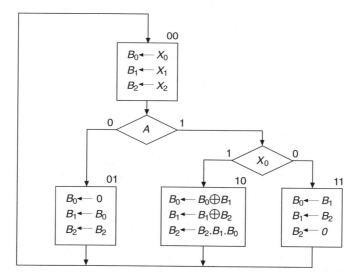

Figure 7.8   ASM chart of partitioned design.

significant bit is tested, so as always to round up. From the ASM chart we can derive next state equations for the controller, either formally or by inspection:

$$S_0^+ = \overline{S_0}.\overline{S_1}.(\overline{A} + \overline{X_0})$$
$$S_1^+ = \overline{S_0}.\overline{S_1}.A$$

The datapath part of the design can be implemented using registers for B_0, B_1 and B_2 and multiplexers, controlled by S_0 and S_1, to select the inputs to the registers, as shown in Figure 7.9. It is also possible to implement the input logic using standard gates and thus to simplify the logic slightly.

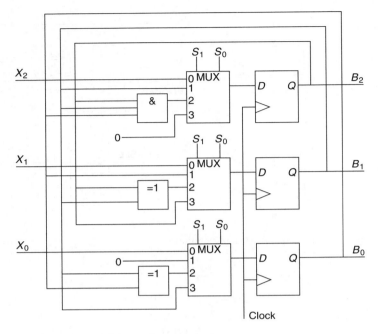

Figure 7.9 Implementation of datapath.

Instructions

Before looking at how a very simple microprocessor can be constructed, we will examine the interface between hardware and software. This is not a book on computer architecture – many such books exist – so the concepts presented here are deliberately simplified.

When a computer program written in, say, C is compiled, the complex expressions of the high-level language can be broken down into a sequence of simple assembler instructions. These assembler instructions can then be directly translated into machine code instructions.[1] These machine code instructions are sets of, say, 8, 16 or 32 bits. It is the interpretation of these bits that is of interest here.

Let us compile the expression

```
a = b + c;
```

to a sequence of assembly code instructions:

```
LOAD b
ADD c
STORE a
```

---

[1]In fact, most compilers would compile directly to machine code. For the purposes of this discussion, it is easier to think in terms of assembler instructions.

The exact interpretation of these assembler instructions will be explained in the next section. If the microprocessor has eight bits, the opcode (LOAD, STORE, etc.) might require three bits, while the operand (a, b, etc.) would take five bits. This allows for eight opcodes and 32 addresses (this is a *very* basic microprocessor). Hence, we might find that the instructions translate as follows:

```
LOAD  b  00000001
ADD   c  01000010
STORE a  00100011
```

i.e. LOAD, ADD and STORE translate to 000, 010 and 001, respectively, while a, b and c are data at addresses 00011, 00001 and 00010, respectively.

Within the microprocessor there is the datapath/controller partition described in the last section. The controller (often known as a sequencer in this context) is a state machine. In the simplest case, the bits of the opcode part of the instruction are inputs to the controller, in the same way that A and X_0 are inputs to the controller of Figure 7.8. Alternatively, the opcode may be decoded (using a decoder implemented in ROM) to generate a larger set of inputs to the controller. The decoder pattern stored in the ROM is known as *microcode*.

The instructions shown above consist of an opcode and an address. The data to be operated upon must be subsequently obtained from the memory addresses given in the instruction. This is known as *direct* addressing. Other addressing modes are possible. Suppose we wish to compile

```
a = b + 5;
```

This translates to:

```
LOAD b
ADD  5
STORE a
```

How do we know that the 5 in the ADD instruction means the value '5' and not the data stored at address 5? In assembler language, we would normally use a special notation, e.g. 'ADD #5', where the '#' indicates to the assembler that the following value is to be interpreted as a value and not as an address. This form of addressing is known as *immediate* mode addressing.

When the microprocessor executes an immediate mode instruction, different parts of the datapath are used compared with those activated by a direct mode instruction. Hence the controller goes through a different sequence of states, and thus the opcodes for an immediate mode ADD and a direct mode ADD must be different. In other words, from the point of view of the microprocessor, instructions with different addressing modes are treated as totally distinct instructions and have different opcodes.

7.4 A simple microprocessor

Using the idea of partitioning a design into a controller and datapath, we will now show how a very basic microprocessor can be designed. We want to be able to execute simple direct mode instructions such as those described in the previous section. Let us first consider the components of the datapath that we need. In order to simplify the routing of data around the microprocessor, we will assume the existence of a single bus. More advanced designs would have two or three buses, but one bus is sufficient for our needs. For simplicity we shall assume that the bus and all the datapath components are eight bits wide, although we shall make the VHDL model, in the next section, parameterizable. Because the single bus may be driven by a number of different components, each of those components will use three-state buffers to ensure that only one component is attempting to put valid data on the bus at a time. We will keep the design fully synchronous, with a single clock driving all sequential blocks. We will also include a single asynchronous reset to initialize all sequential blocks. A block diagram of the microprocessor is shown in Figure 7.10.

The program to be executed by the microprocessor will be held in memory together with any data. Memory such as SRAM is commonly asynchronous, therefore synchronous registers will be included as buffers between the memory and the bus for both the address and data signals. These registers are the Memory Address Register (MAR) and Memory Data Register (MDR).

The Arithmetic and Logic Unit (ALU) performs the arithmetic operations (e.g. ADD). The ALU is a combinational block. The result of an arithmetic operation is held in a register, called the Accumulator (ACC). The inputs to the ALU are the bus and the ACC. The ALU may also have further outputs, or flags, to indicate that the result in the ACC has a particular characteristic, such as being negative. These flags act as inputs to the sequencer.

The various instructions of a program are held sequentially in memory. Therefore the address of the next instruction to be executed needs to be stored. This is done using the Program Counter (PC), which also includes the necessary logic to automatically increment the address held in the PC. If a branch is executed, the program executes out of sequence, so it must also be possible to load a new address into the PC.

Finally, an instruction taken from the memory needs to be stored and acted upon. The Instruction Register (IR) holds the current instruction. The bits corresponding

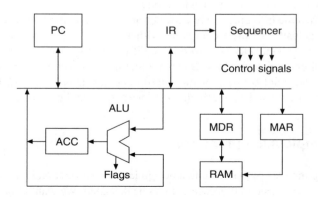

Figure 7.10 Datapath of CPU.

Table 7.1	Control signals of microprocessor.
ACC_bus	Drive bus with contents of ACC (enable three-state output)
load_ACC	Load ACC from bus
PC_bus	Drive bus with contents of PC
load_IR	Load IR from bus
load_MAR	Load MAR from bus
MDR_bus	Drive bus with contents of MDR
load_MDR	Load MDR from bus
ALU_ACC	Load ACC with result from ALU
INC_PC	Increment PC and save the result in PC
Addr_bus	Drive bus with operand part of instruction held in IR
CS	Chip Select. Use contents of MAR to set up memory address
R_NW	Read, Not Write. When false, contents of MDR are stored in memory
ALU_add	Perform an add operation in the ALU
ALU_sub	Perform a subtract operation in the ALU

to the opcode are inputs to the sequencer, which is the state machine controlling the overall functioning of the microprocessor.

The sequencer generates a number of control signals. These determine which components can write to the bus, which registers are loaded from the bus and which ALU operations are performed. The control signals for this microprocessor are listed in Table 7.1.

Figure 7.11 shows the ASM chart of the microprocessor sequencer. Six clock cycles are required to complete each instruction. The execution cycle can be divided into two parts: the *fetch* phase and the *execute* phase. In the first state of the fetch phase, $s0$, the contents of the PC are loaded, via the bus, into MAR. At the same time the address in the PC is incremented by 1. In state $s1$, the CS and R_NW signals are both asserted to read into MDR the contents of the memory at the address given by MAR. In state $s2$, the contents of MDR are transferred to IR via the bus.

In the execute phase, the instruction, now held in IR, is interpreted and executed. In state $s3$, the address part of the instruction, the operand, is copied back to MAR, in anticipation of using it to load or store further data. If the opcode held in IR is STORE, control passes through $s4$ and $s5$, in which the contents of ACC are transferred to MDR, then to be written into memory (at the address previously stored in MAR) when CS is asserted. If the opcode is not STORE, CS and R_NW are asserted in state $s6$, to read data from memory into MDR. If the opcode is LOAD, the contents of MDR are transferred to ACC in state $s7$, otherwise an arithmetic operation is performed by the ALU using the data in ACC and in MDR in state $s8$. The results of this operation are stored in ACC.

The ASM chart of Figure 7.11 shows register transfer operations. In Figure 7.12, the ASM chart shows instead the control signals that are asserted in each state. Either form is valid, although that of Figure 7.11 is more abstract.

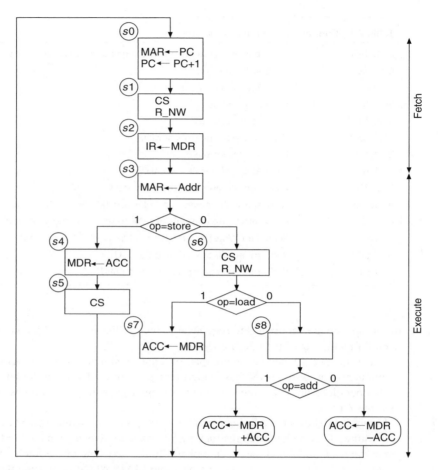

Figure 7.11   ASM chart of microprocessor.

This processor does not include branching. Hence, it is of little use for running programs. Let us extend the microprocessor to include a branch if the result of an arithmetic operation (stored in ACC) is not zero (BNE). The ALU has a *zero flag* which is true if the result it calculates is zero and which is an input to the sequencer. Here, we shall implement this branch instruction in a somewhat unusual manner. All the instructions in this example are direct mode instructions. To implement immediate mode instructions would require significant alteration of the ASM chart. Therefore we will implement a 'direct mode branch'. The operand of a BNE instruction is not the address to which the microprocessor will branch (if the zero flag is true), but the address at which this destination address is stored. Figure 7.13 shows how the lower right corner of the ASM chart can be modified to include this branch. An additional control signal has to be included: load_PC, to load the PC from the bus.

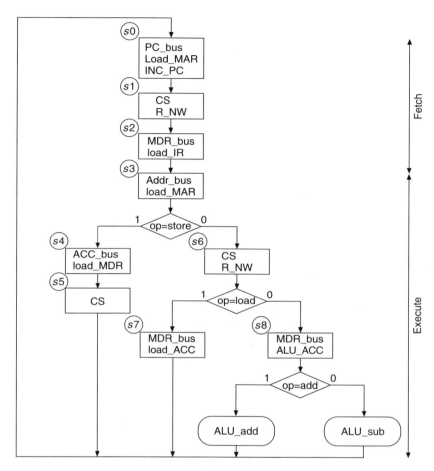

Figure 7.12 Alternative form of microprocessor ASM chart.

7.5 VHDL model of a simple microprocessor

The following VHDL units model the microprocessor described in the previous section. The entire model, including a basic testbench, runs to around 320 lines of code. The model is synthesizable (with one minor modification) and so could be implemented on an FPGA.

The first unit is a set of definitions, contained in a **package**. The definitions in the package are public and may be used in any unit that references the package. The opcodes are defined as an enumerated type. Bit patterns corresponding to the opcodes are defined in the package body, which is private. Therefore the package contains two conversion functions for translating between the abstract opcodes and the bit patterns: slv2op and op2slv. The size of the bus and the number of bits in the opcode are defined by constants. The use of this package means that the size of the CPU and the actual opcodes can be changed without altering any other part of the model. This is important in maintaining the modularity of the design.

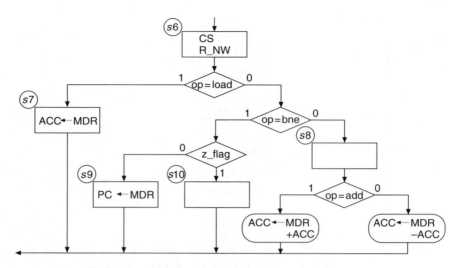

Figure 7.13   Modification of ASM chart to include branching.

```
library IEEE;
use IEEE.std_logic_1164.all;

package cpu_defs is
  type opcode is (load, store, add, sub, bne);
  constant word_w : NATURAL := 8;-- no. of bits for bus
  constant op_w : NATURAL := 3;-- no. of bits for opcode
  constant rfill : std_logic_vector(op_w - 1 downto 0)
                   := (others => '0');
                   -- padding for address
  function slv2op (slv : in std_logic_vector) return
                opcode;
  function op2slv (op : in opcode) return
                std_logic_vector;
end package cpu_defs;

package body cpu_defs is
  type optable is array (opcode)
      of std_logic_vector(op_w - 1 downto 0);
  constant trans_table : optable
            := ("000", "001", "010", "011", "100");
  function op2slv (op : in opcode) return
                std_logic_vector is
  begin
    return trans_table(op);
  end function op2slv;
  function slv2op (slv : in std_logic_vector) return
                opcode is
    variable transop : opcode;
```

```
   begin
-- This is how it should be done, but some synthesis
-- tools may not support this.
   for i in opcode loop
      if slv = trans_table(i) then
         transop := i;
      end if;
   end loop;
-- This is a less elegant method! If the definitions
-- of opcode and/or trans_table are changed, this
-- code also has to be changed. There is therefore
-- potential for inconsistency.
--    case slv is
--      when "000" => transop := load;
--      when "001" => transop := store;
--      when "010" => transop := add;
--      when "011" => transop := sub;
--      when "100" => transop := bne;
--      when others => null;
--    end case;
      return transop;
   end function slv2op;
end package body cpu_defs;
```

The controller or sequencer is described above by an ASM chart. Therefore the VHDL description also takes the form of a state machine. The inputs to the state machine are the clock, reset, an opcode and the zero flag from the accumulator. The outputs are the control signals of Table 7.1. Notice that the two-process model is used. Notice, too, that all the output signals are given a default value at the start of the next state and output logic process.

```
library IEEE;
use IEEE.std_logic_1164.all;
use WORK.cpu_defs.all;

entity sequencer is
  port (clock, reset, z_flag : in std_logic;
        op : in opcode;
        ACC_bus, load_ACC, PC_bus, load_PC,
        load_IR, load_MAR, MDR_bus, load_MDR,
        ALU_ACC, ALU_add, ALU_sub, INC_PC,
        Addr_bus, CS, R_NW : out std_logic);
end entity sequencer;

architecture rtl of sequencer is
  type state is (s0, s1, s2, s3, s4, s5, s6, s7, s8, s9, s10);
  signal present_state, next_state : state;
begin
  seq : process (clock, reset) is
```

```vhdl
begin
  if reset = '1' then
    present_state <= s0;
  elsif rising_edge(clock) then
    present_state <= next_state;
  end if;
end process seq;
com : process (present_state, op, z_flag) is
begin
  -- reset all the control signals to default
  ACC_bus <= '0';
  load_ACC <= '0';
  PC_bus <= '0';
  load_PC <= '0';
  load_IR <= '0';
  load_MAR <= '0';
  MDR_bus <= '0';
  load_MDR <= '0';
  ALU_ACC <= '0';
  ALU_add <= '0';
  ALU_sub <= '0';
  INC_PC <= '0';
  Addr_bus <= '0';
  CS <= '0';
  R_NW <= '0';
  case present_state is
    when s0 =>
      PC_bus <= '1';
      load_MAR <= '1';
      INC_PC <= '1';
      load_PC <= '1';
      next_state <= s1;
    when s1 =>
      CS <= '1';
      R_NW <= '1';
      next_state <= s2;
    when s2 =>
      MDR_bus <= '1';
      load_IR <= '1';
      next_state <= s3;
    when s3 =>
      Addr_bus <= '1';
      load_MAR <= '1';
      if op = store then
        next_state <= s4;
      else
```

```vhdl
                next_state <= s6;
              end if;
          when s4 =>
            ACC_bus <= '1';
            load_MDR <= '1';
            next_state <= s5;
          when s5 =>
            CS <= '1';
            next_state <= s0;
          when s6 =>
            CS <= '1';
            R_NW <= '1';
            if op = load then
              next_state <= s7;
            elsif op = bne then
              if z_flag = '0' then
                next_state <= s9;
              else
                next_state <= s10;
              end if;
            else
              next_state <= s8;
            end if;
          when s7 =>
            MDR_bus <= '1';
            load_ACC <= '1';
            next_state <= s0;
          when s8 =>
            MDR_bus <= '1';
            ALU_ACC <= '1';
            load_ACC <= '1';
            if op = add then
              ALU_add <= '1';
            elsif op = sub then
              ALU_sub <= '1';
            end if;
            next_state <= s0;
          when s9 =>
            MDR_bus <= '1';
            load_PC <= '1';
            next_state <= s0;
          when s10 =>
            next_state <= s0;
        end case;
    end process com;
end architecture rtl;
```

The datapath side of the design, as shown in Figure 7.10, has been described in four parts. Each of these parts is similar to the type of sequential building block described in the last chapter. The system bus is described as a bidirectional port in each of the following four modules. A concurrent assignment sets a high impedance state onto the bus unless the appropriate output enable signal is set. In the **port** declarations, sysbus has **inout** mode. The first module models the ALU and Accumulator.

```vhdl
library IEEE;
use IEEE.std_logic_1164.all, IEEE.numeric_std.all;
use WORK.cpu_defs.all;

entity ALU is
  port (clock, reset : in std_logic;
        ACC_bus, load_ACC, ALU_ACC, ALU_add,
        ALU_sub : in std_logic;
        sysbus : inout std_logic_vector
                          (word_w - 1 downto 0);
        z_flag : out std_logic);
end entity ALU;

architecture rtl of ALU is
  signal acc : unsigned(word_w - 1 downto 0);
  constant zero : unsigned(word_w - 1 downto 0) :=
                          (others => '0');
begin
  sysbus <= std_logic_vector(acc) when ACC_bus = '1'
            else (others => 'Z');
  z_flag <= '1' when acc = zero else '0';
  process (clock, reset) is
  begin
    if reset = '1' then
      acc <= (others => '0');
    elsif rising_edge(clock) then
      if load_ACC = '1' then
        if ALU_ACC = '1' then
          if ALU_add = '1' then
            acc <= acc + unsigned(sysbus);
          elsif ALU_sub = '1' then
            acc <= acc - unsigned(sysbus);
          end if;
        else
          acc <= unsigned(sysbus);
        end if;
      end if;
    end if;
  end process;
end architecture rtl;
```

The program counter is similar in structure to the ALU and Accumulator.

```vhdl
library IEEE;
use IEEE.std_logic_1164.all, IEEE.numeric_std.all;
use WORK.cpu_defs.all;

entity PC is
  port (clock, reset : in std_logic;
        PC_bus, load_PC, INC_PC : in std_logic;
        sysbus : inout std_logic_vector
                      (word_w - 1 downto 0));
end entity PC;

architecture rtl of PC is
  signal count : unsigned(word_w - op_w - 1 downto 0);
begin
  sysbus <= rfill & std_logic_vector(count)
              when PC_bus = '1' else (others => 'Z');
  process (clock, reset) is
  begin
    if reset = '1' then
      count <= (others => '0');
    elsif rising_edge(clock) then
      if load_PC = '1' then
        if INC_PC = '1' then
          count <= count + 1;
        else
          count <= unsigned(
                    sysbus(word_w - op_w - 1 downto 0));
        end if;
      end if;
    end if;
  end process;
end architecture rtl;
```

The instruction register is basically an enabled register. The opcode is decoded for input to the sequencer.

```vhdl
library IEEE;
use IEEE.std_logic_1164.all;
use WORK.cpu_defs.all;

entity IR is
  port (clock, reset : in std_logic;
        Addr_bus, load_IR : in std_logic;
        op : out opcode;
        sysbus : inout std_logic_vector
                      (word_w - 1 downto 0));
end entity IR;
```

```
architecture rtl of IR is
  signal instr_reg: std_logic_vector(word_w - 1 downto 0);
begin
  sysbus <= rfill & instr_reg(word_w - op_w - 1 downto 0)
              when Addr_bus = '1' else (others => 'Z');
  op <= slv2op(instr_reg(word_w - 1 downto word_w - op_w));
  process (clock, reset) is
  begin
    if reset = '1' then
      instr_reg <= (others => '0');
    elsif rising_edge(clock) then
      if load_IR = '1' then
        instr_reg <= sysbus;
      end if;
    end if;
  end process;
end architecture rtl;
```

The memory module is, again, very similar to the static RAM of the last chapter. A short program has been loaded in the RAM, using constant declarations.

```
library IEEE;
use IEEE.std_logic_1164.all, IEEE.numeric_std.all;
use WORK.cpu_defs.all;

entity RAM is
  port (clock, reset : in std_logic;
        MDR_bus, load_MDR, load_MAR, CS,
        R_NW : in std_logic;
        sysbus : inout std_logic_vector
                        (word_w - 1 downto 0));
end entity RAM;

architecture rtl of RAM is
  signal mdr : std_logic_vector(word_w - 1 downto 0);
  signal mar : unsigned(word_w - op_w - 1 downto 0);
begin
  sysbus <= mdr when MDR_bus = '1' else (others => 'Z');
  process (clock, reset) is
    type mem_array is array (0 to 2**(word_w - op_w) - 1)
        of std_logic_vector(word_w - 1 downto 0);
    variable mem : mem_array;
    constant prog : mem_array := (
      0 => op2slv(load) &
        std_logic_vector(to_unsigned(4, word_w - op_w)),
      1 => op2slv(add) &
        std_logic_vector(to_unsigned(5, word_w - op_w)),
```

```vhdl
        2 => op2slv(store) &
          std_logic_vector(to_unsigned(6, word_w - op_w)),
        3 => op2slv(bne) &
          std_logic_vector(to_unsigned(7, word_w - op_w)),
        4 => std_logic_vector(to_unsigned(2, word_w)),
        5 => std_logic_vector(to_unsigned(2, word_w)),
        others => (others => '0'));
  begin
      if reset = '1' then
        mdr <= (others => '0');
        mar <= (others => '0');
        mem := prog;
      elsif rising_edge(clock) then
        if load_MAR = '1' then
          mar <= unsigned(sysbus(word_w - op_w - 1 downto 0));
        elsif load_MDR = '1' then
          mdr <= sysbus;
        elsif CS = '1' then
          if R_NW = '1' then
            mdr <= mem(to_integer(mar));
          else
            mem(to_integer(mar)) := mdr;
          end if;
        end if;
      end if;
  end process;
end architecture rtl;
```

The various parts of the microprocessor can now be pulled together by instantiating them as components.

```vhdl
library IEEE;
use IEEE.std_logic_1164.all;
use WORK.cpu_defs.all;

entity CPU is
  port (clock, reset : in std_logic;
        sysbus : inout std_logic_vector
                        (word_w - 1 downto 0));
end entity CPU;

architecture top of CPU is
  signal ACC_bus, load_ACC, PC_bus, load_PC,
         load_IR, load_MAR, MDR_bus, load_MDR,
         ALU_ACC, ALU_add, ALU_sub, INC_PC,
         Addr_bus, CS, R_NW, z_flag : std_logic;
  signal op : opcode;
```

```
begin
s1: entity WORK.sequencer port map (clock, reset,
                            z_flag, op, ACC_bus, load_ACC,
                            PC_bus, load_PC, load_IR,
                            load_MAR, MDR_bus, load_MDR,
                            ALU_ACC, ALU_add, ALU_sub,
                            INC_PC, Addr_bus, CS, R_NW);
i1: entity WORK.IR port map (clock, reset, Addr_bus,
                            load_IR, op, sysbus);
p1: entity WORK.PC port map (clock, reset, PC_bus,
                            load_PC, INC_PC, sysbus);
a1: entity WORK.ALU port map (clock, reset, ACC_bus,
                            load_ACC, ALU_ACC, ALU_add,
                            ALU_sub, sysbus, z_flag);
r1: entity WORK.RAM port map (clock, reset, MDR_bus,
                            load_MDR, load_MAR, CS, R_NW,
                            sysbus);
end architecture top;
```

The following piece of VHDL generates a clock and reset signal to allow the program defined in the RAM module to be executed. Obviously, this testbench would not be synthesized.

```
library IEEE;
use IEEE.std_logic_1164.all;
use WORK.cpu_defs.all;

entity testcpu is
end entity testcpu;

architecture tb of testcpu is
  signal clock, reset : std_logic := '0';
  signal sysbus : std_logic_vector(word_w - 1 downto 0);
begin
  c1 : entity WORK.CPU port map (clock, reset, sysbus);
  reset <= '1' after 1 ns, '0' after 2 ns;
  clock <= not clock after 10 ns;
end architecture tb;
```

Summary

In this chapter we have looked at linked ASM charts and at splitting a design between a controller, which is designed using formal sequential design methods, and a datapath that consists of standard building blocks. The example of a simple CPU has been used to illustrate this partitioning. The VHDL model can be both simulated and synthesized.

Further reading

Formal techniques exist for partitioning state machines. These are described in Unger. The controller/datapath model is used in a number of high-level synthesis tools; see, for example, de Micheli. The CPU model is based on an example in Maccabe.

Exercises

7.1 Any synchronous sequential system can be described by a single ASM chart. Why then might it be desirable to partition a design? Describe a general partitioning scheme.

7.2 A counter is to be used as a delay for a simple controller, to generate a ready signal, 10 clock cycles after a start signal has been asserted. Show how the interaction between the controller and the counter can be represented in ASM chart notation.

7.3 A microprocessor has a number of addressing modes. Describe the immediate and direct addressing modes.

7.4 What structures are needed in a microprocessor to implement a 'branch if negative' instruction? Describe the register transfer operations that would occur in the execution of such an instruction and show the sequence of events on a timing diagram.

7.5 The ASM chart of Figures 7.11 and 7.13 implements a branch instruction with a direct mode operand. Modify the ASM chart to show how the microprocessor could branch to an address given by an immediate mode operand.

7.6 Modify the VHDL model of the microprocessor to implement an immediate mode 'branch if not equal to zero'.

VHDL simulation

VHDL is a language for describing digital systems. To verify that a model is correct, a simulator may be used to animate the model. In the first section of this chapter, the principles of digital simulation are described. The specifics of VHDL simulation and techniques to improve simulation efficiency are then discussed. Finally, file handling is described.

8.1 Event-driven simulation

VHDL is a language for describing digital systems; therefore it should be no surprise that standard event-driven logic simulation algorithms are used. Such algorithms are most easily described in terms of the simulation of structural models. Behavioural models are evaluated in much the same way, where a process can be thought of as an element.

The objective of event-driven simulation is to minimize the work done by the simulator. Therefore the state of the circuit is evaluated only when a change occurs in the circuit. It is possible to predict when the output of an element might change because we know that such a change can occur only after an input changes. If we monitor only the inputs to elements, we can know only that an output might change; the logic function of the element determines whether or not a change actually occurs. As we also know the delays through the element, we know when the output might change. Thus an element needs to be evaluated only when it is known that its output might change but not otherwise. Nevertheless, even by predicting a possible change, it is necessary to re-evaluate elements only when the possible changes occur. By following the possible *events* through the circuit we can minimize the computation done by the simulator. Only elements that change need to be evaluated; anything that is not changing is ignored.

The delays through elements are defined in terms of integer times. The units of time might be nanoseconds or picoseconds. As the time is incremented in discrete intervals, it is likely that, for any reasonably large circuit, more than one element will be evaluated at any one time. Equally, there may be times at which no elements are due for evaluation. This implies a form of time step control. As each element is evaluated, any change in its output will cause inputs to further elements to change, and hence the outputs of those elements may subsequently change. Clearly it is necessary to maintain a list of which signals change and when. An *event* is therefore a new value of a signal, together with the time at which the signal changes. If the event list is ordered in time, it should be easy to add new events at the correct place in the future.

A suitable data structure for the event list is shown in Figure 8.1. When an event is predicted, it is added to the list of events at the predicted time. When an event is processed it is removed from the list. When all the events at a particular time have been processed, that time can be removed.

This list manipulation is relatively easy to do in a block-structured programming language, such as C, although adding new times to the middle of a list can be awkward.

An element can be scheduled for processing when its inputs are known to change. For example, consider an AND gate with a 4 ns delay. When the signal at one input changes it can be predicted whether the signal at the output changes depending on the state of the other inputs of the gate. If the output does change, the output event can be scheduled 4 ns later. The algorithm can be written as shown in pseudo-C in Figure 8.2.

An event is scheduled only if the new value is different from the value that has previously been scheduled for that signal. If two or more events occur on input signals to

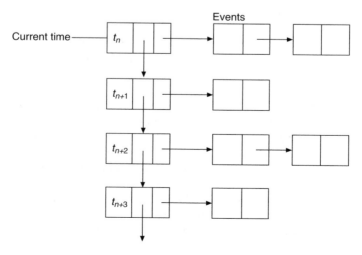

Figure 8.1   Event list.

```
for (i = each event at current_time)
  {
  update_node(i);
  for (j = each gate on the fanout list of i)
    {
    update input values of j;
    evaluate(j);
    if (new_value(j) ≠ last_scheduled_value(j) ) {
      schedule new_value(j) at
        (current_time + delay(j) );
      last_scheduled_value(j) = new_value(j);
      }
    }
  }
```

Figure 8.2 Single-pass event scheduler.

an element, more than one event may be scheduled for an output signal. It is thus important to know that the new value is different not merely from the present value but also from a value that might already be scheduled to be set in the future. This algorithm therefore has a disadvantage as it stands, because an element is evaluated whenever an event occurs at an input. It is quite possible that two events might be scheduled for the same gate at the same time. This could lead to a zero-width spike being scheduled one delay later. Even worse, if the delays for rising and falling output differ, the presence or absence of an output pulse would depend on the order in which the input events were processed.

In VHDL terminology, an input change that might lead to an output change is known as a *transaction*. Only if an output changes is this transaction scheduled as an event.

If zero-width pulses are to be suppressed, they can be considered as a special case of the inertial delay model. The previous algorithm can be extended to include pulse cancellation if a pulse is less than the permitted minimum width, as shown in Figure 8.3. This model assumes two-state logic. If an event is predicted at a time less than the inertial delay after the previous event for that node, this new event is not set and the previous event is also removed. If more than two-state logic is used, the meaning of an inertial delay and hence of a cancelled event must be thought about carefully. In order to cancel an event, it is necessary to search through the event lists. Event cancellation is therefore best avoided, if possible.

One further problem exists with the selective trace algorithm. A gate with a zero delay would cause an event to be scheduled at the current simulation time if an input changes. Thus, while events are being processed at the current time, new events are being added to the end of the event list. There is clearly the potential here for an infinite loop to be generated, where the simulation never advances beyond the current simulation time. In practice, the only way to avoid this problem is to count the number of events at a time point, and if they exceed some arbitrary limit to terminate the simulation.

We have already noted that the presence or absence of zero-width pulses can be dependent upon the order of evaluation of events at a time point. Consider the circuit of Figure 8.4. If both gates have a zero delay and the input changes from 0 to 1 as shown, a zero-width pulse may be generated.

```
for (i = each event at current_time)
  {
  update_node(i);
  for (j = each gate on the fanout list of i)
    {
    update input values of j;
    evaluate(j);
    if (new_value(j) ≠ last_scheduled_value(j) )
    {
    schedule_time = current_time + delay(j);
    if (schedule_time ≤ last_scheduled_time(j) + inertial_delay(j))
      {
      cancel_event at last_scheduled_time(j);
      }
    else {
      schedule new_value(j) at schedule_time;
      }
    last_scheduled_value(j) = new_value(j);
    last_scheduled_time(j) = schedule_time;
    }
  }
}
```

Figure 8.3   Single-pass event scheduler with inertial delay cancellation.

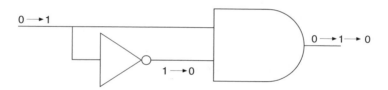

Figure 8.4   Circuit with zero delays.

If the AND gate is evaluated first, both inputs will appear to be at logic 1, so the output will become 1. The inverter is evaluated next, causing the other AND input to change. The AND gate is evaluated again and the output changes back to 0. On the other hand, if the inverter is evaluated first, both inputs to the AND gate will appear to change simultaneously when it is evaluated, and no pulse is generated. Although it is obvious here that the inverter should be evaluated first, this is not always the case and we must assume that the order of evaluation of gates is effectively arbitrary.

This arbitrariness can be avoided by using *delta delays*. A delta delay can be thought of an infinitesimal unit of time, as shown in Figure 8.5. Zero delays are modelled as delta delays, so that any events generated at the current time with zero delays are scheduled to occur one delta delay later. In the above example, both gates would be evaluated at the current time, and their outputs would be scheduled one delta delay later. The AND gate would then be evaluated again at the current time plus one delta. Thus a circuit is always evaluated in exactly the same way; in this case we always get

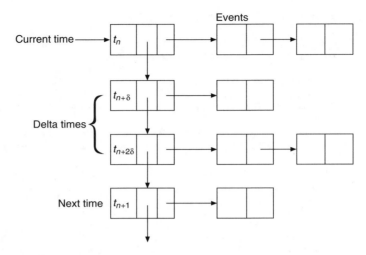

Figure 8.5   Event list with delta delays.

a zero-width pulse generated. Moreover, any simulator that implements delta delays will evaluate the same circuit in the same way, so we can directly compare simulators. It should be noted, of course, that the use of delta delays causes new delta times to be inserted into the event list, but it is not possible for a series of delta times to add up to a real simulation time step.

8.2   Simulation of VHDL models

8.2.1   Simulation time

We have already seen how transport and inertial delays may be modelled in VHDL. The concept of delta delays is used in VHDL simulators to ensure that concurrent statements are always executed consistently. Several VHDL functions and attributes exist to allow interaction with a simulator, either for model writing, or perhaps more usefully, for testbench modelling.

A function now returns the current simulated time. TIME is a predefined VHDL physical type:

```
type TIME is range <implementation defined>
  units
    FS; -- femtosecond
    PS = 1000 FS; -- picosecond
    NS = 1000 PS; -- nanosecond
    US = 1000 NS; -- microsecond
    MS = 1000 US; -- millisecond
    SEC = 1000 MS; -- second
```

```
    MIN = 60 SEC; -- minute
    HR = 60 MIN; -- hour
end units TIME;
```

(Note that any physical type can be defined with units in this way.)

A signal, s, has several attributes that may be considered as relevant to simulation models:

s'EVENT – TRUE if an event occurred on that signal during the current simulation delta cycle

s'ACTIVE – TRUE if a transaction occurred during the current simulation delta cycle

s'LAST_EVENT – returns the time since the last event

s'LAST_VALUE – returns the value of the signal at its last event

s'LAST_ACTIVE – returns the time since the last transaction

s'TRANSACTION – creates a signal of type BIT that toggles its value whenever a transaction (or event) occurs on the signal

s'DELAYED[(<TIME>)] – creates a signal of the same type as s, delayed by the specified time

s'STABLE[(<TIME>)] – creates a signal of type Boolean that is true when the signal has had no events for the specified time

s'QUIET[(<TIME>)] – creates a signal of type Boolean that is true when the signal has had no transactions or events for the specified time.

Attributes 'TRANSACTION, 'DELAYED, 'STABLE and 'QUIET create *implicit* signals. All other signals are *explicit*. Signals have a *driving* value and an *effective* value. The driving value of a signal is one of the following: the default value of the signal type if the signal has no source; the value of the driver source or port (if the signal has a single driver); or the resolved signal value (if the signal has multiple drivers). The effective value of a signal is normally the same as the driving value, but may be changed when the signal is propagated through conversion functions or type conversions.

8.2.2   Compilation and elaboration

Before simulation, or indeed synthesis, a VHDL description has to be compiled. In order for the compilation of a unit to succeed, all the definitions used in that unit must have been compiled first. For example, an entity description must be compiled before any of its associated architectures. Similarly, a package header must be compiled before a package body. If an entity uses a package, the package header must be compiled before the entity, although the package body, containing hidden definitions, may be compiled later. For a project such as the CPU of the previous chapter, the order of compilation is therefore important. The cpu_defs package must be compiled first.

In general, configuration units can be compiled last, as they define the *binding*, in other words which architecture to use for a given entity. It should, however, be noted that some compilers are 'over-enthusiastic' in that they try to resolve the binding as the compilation proceeds. In the absence of any configuration units or configuration statements, a *default binding* will be used. On meeting a component declaration, the compiler may look to see whether the corresponding entity has been declared. This can lead to confusing warning messages.

Elaboration is the process of building a complete and consistent simulation model from the various compiled units of a design. The rules for elaboration are described over about 11 pages of the VHDL LRM and it would therefore be inappropriate to attempt to summarize them here. The basic principle, however, is that all parts of the design must be fully specified before simulation or synthesis can proceed. Thus, for example, types must be fully defined, architectures must exist for all components, and array and iteration bounds must be defined.

8.2.3   VHDL simulation cycle

Once compilation and elaboration are complete, the simulation can begin. The simulator itself can be thought of as a kernel process containing the event list and that updates signals. The VHDL model has a number of other processes – those explicitly defined as processes and concurrent statements, each of which can be thought of as a process. A process is said to be *passive* if it does not contain any signal assignments. Processes (including concurrent statements) can be declared to be **postponed**. Postponed processes are executed at the end of each simulation cycle (see below) and must therefore not give rise to further events at the current simulation time.

In Chapter 4, we introduced the concept of a *driver*. If a signal is assigned a value in a process, that signal is said to have a driver in that process. A driver is represented by a *projected output waveform*, which consists of a sequence of transactions. As the simulation time advances, each transaction in turn becomes the value of the driver. At the time that a driver takes a new value, the driver is *active*. If one or more drivers of a signal is active, that signal is active.

VHDL simulation consists of two parts: an *initialization phase*, followed by repeated executions of the *simulation cycle*. The simulation starts at time 0 ns. The current simulation time is T_c. The next simulation time is T_n. The maximum possible simulation time is implementation dependent, but is denoted by TIME'HIGH. The value of T_n is calculated as the earliest occurring of:

1. TIME'HIGH
2. The next time at which any driver becomes active
3. The next time at which any process resumes.

The initialization phase therefore consists of the following steps.

1. Compute the driving and effective values of all explicit signals. Signals are assumed to have held their effective values for an infinite amount of time before the start of the simulation.
2. Set the values of the 'QUIET and 'STABLE implicit signals to TRUE and the values of 'DELAYED signals to the initial values of the signals that are delayed. The initial value of 'TRANSACTION implicit signals is not defined.

3. Each non-postponed process is executed until it suspends. A process suspends at the first **wait** statement or at the end of the process if there are no **wait** statements. Note that the order of execution of process is *not* determinate.

4. Each postponed process is executed until it suspends.

5. The time, T_n, of the first simulation cycle is determined as shown above.

Each simulation cycle executes as follows:

1. T_c is set equal to T_n. The simulation completes when T_c is equal to TIME'HIGH, and there are no further active drivers or processes to be resumed.

2. Each active explicit signal is updated. Each implicit signal is updated. Both these steps may cause new events to occur, which are scheduled either at a future time, or one delta cycle ahead.

3. If an event occurs on a signal in this simulation cycle and a process is sensitive to that signal, the process resumes.

4. Each non-postponed process that resumes in this simulation cycle is executed until it suspends.

5. The time of the next simulation cycle, T_n, is determined, as above. If T_n is equal to T_c, the next cycle is a delta cycle.

6. If the next cycle is *not* a delta cycle (i.e. T_n is greater than T_c), each postponed process that resumes in this simulation cycle is executed until it suspends. T_n is determined again, as above. The **postponed** processes must not cause another delta cycle to occur.

8.3 Simulation modelling issues

Clearly, a VHDL model should be correct, in the sense that it accurately reflects the intentions of the user. There are several aspects, specific to simulation, which should be noted.

- A process should have a complete list of relevant signals in its sensitivity list. A process is activated only when one of these signals has an event on it, so an incomplete list will result in the process not being evaluated correctly. Equally, do not include irrelevant signals as the process will be activated unnecessarily.

- Minimize the number of processes.

- There is a significant difference between variable and signal assignments in processes. A variable assignment takes effect *immediately*. Signal assignments (without delays) take effect only at the end of the process, or at the next **wait** statement. Thus the result of a signal assignment is available only in the next delta cycle.

- Use variables if possible, rather than signals. Signals have histories and are therefore large, complicated data structures.

- Use 'EVENT rather than 'STABLE. The latter creates a signal. Use rising_edge and falling_edge functions so that only true transitions are modelled.

- Processes, signal assignments, procedure calls and assertions can be declared to be **postponed**. This means that they are executed after the last delta cycle at a time point. Therefore they must not cause a further delta cycle.

- Normal programming efficiency comments apply – e.g. take invariant assignments out of loops.

- It is possible to do conditional compilation in VHDL (cf. #ifdef in C). An **if . . . generate** statement will create the components or processes inside it only if the condition is true.

- Data types such as integers are likely to simulate faster than bit vectors, because they map directly onto standard data types in C or whatever language the simulator was written in.

8.4 File operations

VHDL models may read and write data from and to files. These files can be used to contain test data or to store results for further analysis. File handling is one of the main causes of portability problems in programs and HDL models. For this reason, it is strongly recommended that only text files are used and further that characters in files are limited to printing characters only.

Here is an example of a non-portable file declaration, using the 1987 standard:

```
type CharFile is file of character;
file DataFile : CharFile is in "data.dat";
```

This is a better way of doing the same thing, using the 1993 and 2002 standards:

```
use STD.textio.all;
file DataFile : text open READ_MODE is "data.dat";
```

The package STD.textio defines the text type as

```
type text is file of STRING;
```

The declaration of a file type implicitly defines the procedures file_open, file_close, read and write and the function endfile. Text is best read and written via a line buffer, so two further procedures, readline and writeline, are defined in STD.textio.

The following is a testbench for the NBitAdder of Section 4.5.2 that uses all these subroutines. Note the conversions between bits and standard logic values. Notice, too, the use of a **wait for** statement to allow time to elapse between applying new signal values.

```
library IEEE;
use IEEE.std_logic_1164.all, STD.textio.all;

entity tb is
end entity tb;
```

```
architecture fileio of tb is
  file vectors : text;
  file results : text;
  constant N : NATURAL := 4;
  signal X, Y, Z : std_logic_vector((N-1) downto 0)
                 := (others => '0');
  signal ci, co: std_logic := '0';
begin
  a1: entity WORK.NBitAdder port map (X, Y, ci, Z, co);
  p1: process is
    variable ILine, OLine : Line;
    variable X_in, Y_in, Z_out : BIT_VECTOR((N-1) downto 0);
    variable ci_in, co_out : BIT;
    variable ch : CHARACTER;
  begin
    file_open(vectors, "vectors.txt", READ_MODE);
    file_open(results, "results.txt", WRITE_MODE);
    while not endfile(vectors) loop
      readline(vectors, ILine);
      read(ILine, X_in);
      read(ILine, ch);
      read(ILine, Y_in);
      read(ILine, ch);
      read(ILine, ci_in);
      X <= to_stdlogicvector(X_in);
      Y <= to_stdlogicvector(Y_in);
      ci <= to_stdulogic(ci_in);
      wait for 60 NS;
      Z_out := to_bitvector(Z);
      co_out := to_bit(co);
      write(OLine, Z_out, right, 5);
      write(OLine, co_out, right, 2);
      writeline(results, OLine);
    end loop;
    file_close(vectors);
    file_close(results);
  end process p1;
end architecture fileio;
```

The testbench reads a file called vectors.txt:

```
0000 0000 0
0000 0001 0
1111 1111 0
1111 1111 1
```

and generates a file called `results.txt`:

```
0000 0
0001 0
1110 1
1111 1
```

Summary

Event-driven simulation minimizes the work of a logic simulator by evaluating only changes. Delta delays ensure that all VHDL simulators produce the same simulation results. The VHDL simulation cycle is specified as part of the LRM. VHDL models can be written to improve simulation speed. Files provide a way of importing data into testbenches.

Further reading

Logic simulation is described in the books by Miczo and by Abramovici, Breuer and Friedman. The VHDL simulation cycle, together with detailed requirements for the elaboration process, is described in the VHDL LRM.

Exercises

8.1 Explain the mechanism used within a VHDL simulator to ensure that the following fragment of VHDL always gives the same simulation results:

```
b <= not a;
c <= a and b;
```

8.2 Explain what is meant by 'inertial' and 'transport' delays. Give an example of how each would be described in a VHDL model.

8.3 VHDL has three forms of **wait** statement. Explain the operation of each. Which form of **wait** statement is equivalent to a process sensitivity list?

8.4 When the following VHDL model of a multiplexer is simulated, it is found that the wrong input may be selected. Explain why, and show how the correct behaviour can be modelled, using (a) **wait** statements and (b) a variable.

```
library ieee;
use ieee.std_logic_1164.all;

entity mux is
  port (a, b, c : in std_logic;
        z : out std_logic);
end entity mux;
```

```
architecture behave of mux is
  signal sel : integer range 0 to 1;
begin
  m1: process (a, b, c) is
  begin
    sel <= 0;
    if (c = '1') then
      sel <= sel + 1;
    end if;
    case sel is
      when 0 =>
        z <= a;
      when 1 =>
        z <= b;
    end case;
  end process m1;
end architecture behave;
```

VHDL synthesis

....

VHDL was originally designed as a hardware *description* language. In other words, the language was designed to model the behaviour of existing hardware, not to specify the functionality of proposed hardware. Moreover, when the VHDL standard was originally written in 1987, there were no automatic synthesis tools in widespread use. Therefore the meaning of different VHDL constructs in hardware terms was derived some years after the language was standardized. The consequence of this is that parts of VHDL are not suitable for synthesis.

We should define, at this point, what we mean by the term *synthesis*. The long-standing objective of design automation tool development has been to compile high-level descriptions into hardware in much the same way that a computer software program is compiled into machine code.

Figure 9.1 shows a simplified view of the design process. After a specification has been agreed, a design can be partitioned into functional units (architectural design). Each of these functional units is then designed as a synchronous system. The design of these parts can be done by hand, as described in Chapter 5. Thus a state machine is designed by formulating an ASM chart, deriving next state and output equations and implementing these in combinational logic. At this point, the gates and registers of the design can be laid out and wired up on an integrated circuit or programmable logic device.

Figure 9.1 shows how synthesis tools can automate parts of this process. RTL (Register Transfer Level) Synthesis tools take a VHDL description of a design in terms of registers, state machines and combinational logic functions and generate a netlist of gates and library cells. As we will see, the VHDL models described in Chapters 4, 5, 6

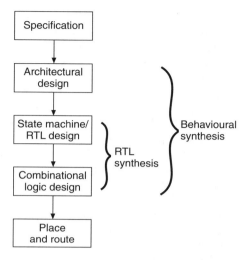

Figure 9.1   High-level design flow.

and 7 are mostly suitable for RTL synthesis. Behavioural synthesis tools, on the other hand, take algorithmic VHDL models and transform them to gates and cells. The user of a behavioural synthesis system would not have to specify clock inputs, for instance, but simply that a particular operation has to be completed within a certain time interval. RTL synthesis tools are gaining widespread acceptance; behavioural synthesis tools are still relatively rare. Although this chapter (and this book) is primarily about RTL synthesis, it is likely that in a few years behavioural synthesis tools will be widely accepted, in a manner analogous to the way that high-level software programming languages such as Java are coming to replace lower-level languages such as C.

The last stage of the synthesis process, place and route, is carried out by separate specialized tools. In the case of programmable logic, the manufacturers of the programmable logic devices often supply these tools.

9.1 RTL synthesis

The functions carried out by an RTL synthesis tool are essentially the same as those described in Chapter 5. The starting point of the synthesis process is a model (in VHDL) of the system we wish to build, described in terms of combinational and sequential building blocks and state machines. Thus we have to know all the inputs and outputs of the system, including the clock and resets. We also have to know the number of states in state machines – in general, RTL synthesis tools do not perform state minimization. From this we can write VHDL models of the parts of our system. In addition, we may wish to define various *constraints*. For instance, we might prefer that a state machine be implemented using a particular form of state encoding. We almost certainly have physical constraints such as the maximum chip size and hence the maximum number of gates in the circuit and the minimum clock frequency at which the system should operate. These constraints are not part of VHDL, in the sense that they do not form part of the simulation model, and are often unique to particular tools, but *may* be included in the VHDL description.

As has already been noted, there are three styles of VHDL: structural, dataflow and behavioural. Structural VHDL takes the form of netlists of components and would not therefore be synthesized any further. Dataflow VHDL – concurrent assignment statements – was originally intended to be used for RTL modelling. Therefore many dataflow constructs are synthesizable. Behavioural (i.e. sequential) VHDL was intended for high-level, algorithmic modelling. As RTL synthesis tools have evolved, it has become clear that certain forms of behavioural VHDL are very suitable for RTL modelling and synthesis. This should be apparent from the VHDL models in earlier chapters. Therefore in this chapter we will highlight ways to describe combinational and sequential hardware for RTL synthesis using behavioural VHDL.

The IEEE standard 1076.6-1999 defines a subset of VHDL for RTL synthesis. The purpose of this standard is to define the minimum subset that can be accepted by *any* synthesis tool. Throughout this book, we have advocated the use of and adherence to various standards. Because IEEE 1076.6 is a minimum subset, it does not, for example, support the 1993 enhancements to VHDL. 1076.6 synthesis tools will almost certainly ignore the syntactic enhancements, but warning messages may be generated.

9.1.1   Non-synthesizable VHDL

Despite the comments above about the distinctions between RTL and behavioural synthesis, in principle most features of VHDL could be translated into hardware. In general, those parts of VHDL that are not synthesizable are constructs in which exact timing is specified and structures whose size is not completely defined. Poorly written VHDL may result in the synthesis of unexpected hardware structures. These will be described later.

The following VHDL constructs are either ignored or rejected by RTL synthesis tools.

- The **after** reserved word and the associated **transport** and **inertial** reserved words and the delay values in such statements are all ignored. Delays in assignments are *simulation* models. A model can be synthesized to meet various *constraints*, but cannot be synthesized to meet some exact timing model. For instance, it is not possible to specify that a gate will have a delay of exactly 5 ns. It is reasonable, on the other hand, to require a synthesis tool to generate a block of combinational logic such that its total delay is less than, say, 20 ns.

- The **wait for** construct is also ignored for the same reason. It suggests that a piece of logic with an exact delay can be built.

- **File** operations suggest the existence of an operating system. Hence file operations cannot be synthesized and would be rejected by a synthesis tool.

- As with the elaboration of VHDL for simulation, the sizes of arrays and other structures must be defined at compile time. Array dimensions may be undefined if they are specified using **generic** parameters. For example, an *n*-bit adder must have the value of n defined before it can be synthesized. Hence generic parameters used to define structure sizes at the topmost level of a hierarchy must have default values.

● In simulation, data structures such as linked lists and trees are elaborated dynamically. Such structures cannot be synthesized because their size is not defined at compile time. Hence pointers, specified by the **access** keyword, are not recognized by synthesis tools.

● Floating-point data types are not inherently *unsynthesizable*, but will be rejected by synthesis tools because they require at least 32 bits, and the hardware required for many operations is too large for most ASICs or FPGAs.

● Initial values of signals and variables will be ignored. Instead, synchronous or asynchronous set and reset signals must be used to initialize flip-flops.

9.1.2   Inferred flip-flops and latches

It is important to appreciate that synthesis tools (like most computer software) are basically stupid. There are no reserved words in VHDL to specify whether a model is combinational or sequential and whether any sequential logic is synchronous or asynchronous. Therefore the fundamental problem with synthesizing VHDL models is to ensure that the hardware produced by the synthesis system is what you really want. One of the most likely 'errors' is the creation of additional flip-flops or latches. Therefore, in this section, we will describe how the existence of flip-flops and latches is inferred.

A flip-flop or latch is synthesized if a signal or variable holds its value over a period of time. In VHDL a signal holds its value until it is given a new value. If a process contains a **wait** statement, any signals (or variables) that have values assigned in that process must hold their values until the **wait** statement completes. Therefore all signals and variables to which assignments are made in a process containing a **wait** must be held in registers. Similarly, a flip-flop or latch is created implicitly if some paths through a process have assignments to a signal or variable while others do not. This typically happens if a **case** statement or an **if** statement is incomplete in the sense that one or more branches do not contain an assignment to a signal while other branches do contain such an assignment, or if the **if** statement does not contain a final **else**.

The term 'flip-flop' refers here to a memory element triggered by an edge of the clock. 'Latch' refers to a level-sensitive device, controlled by some signal other than the clock. Thus a flip-flop would be created if the *edge* of a signal is used in a **wait**, **if** or **case** statement, while a latch would be created if the level value of a signal were used instead.

In principle, therefore, processes containing various edge-triggered and level-sensitive statements could be synthesized. In practice, synthesis tools recognize a small number of fairly simple patterns, as shown in the rest of this section. These examples can act as templates for larger examples. It should be noted that in all these examples, the signal names are not significant to the synthesis tool. Thus a clock signal might be called 'Clock' or 'Clk1' or, with equal validity, 'Data'. Note, however, that good software engineering practice should be applied and *meaningful* identifiers should be used for the benefit of human readers.

Level-sensitive latch

The following example shows the VHDL that would be interpreted to specify a level-sensitive latch by an RTL synthesis tool:

```
p0: process (Ctrl, A) is
begin
  if (Ctrl = '1') then
    Z <= A;
  end if;
end process p0;
```

The process has a sensitivity list containing the signal Ctrl and the signal, A, which is assigned to the output. Therefore the process is executed when either Ctrl or A changes. Z is assigned the value of A if Ctrl has just changed to a 1. While Ctrl is 1, any change in A is transmitted to the output. Otherwise, no assignment to Z is specified. Therefore it may be inferred that Z holds its value, and hence it is inferred that Z is a registered signal. This inference can be avoided if the **else** clause is included:

```
p0: process (Ctrl, A) is
begin
  if (Ctrl = '1') then
    Z <= A;
  else
    Z <= '0';
  end if;
end process p0;
```

The value of Z is therefore Ctrl **and** A. **Case** statements are interpreted in a similar manner.

```
p0: process (Sel, A, B) is
begin
  case Sel is
    when "00" => Y <= A;
    when "10" => Y <= B;
    when others => null;
  end case;
end process p0;
```

The **when others** clause covers the patterns '01' and '11'. If it were omitted, the **case** statement would be incorrect and hence would cause a compilation error. When Sel is one of these two patterns, Y is assumed to hold its value. Hence the circuit of Figure 9.2 is synthesized. (If all the values of Sel were listed in **when** clauses, the **others** statement would not be needed. Some books suggest that if an enumerated

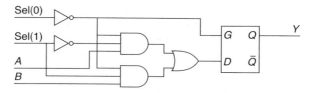

Figure 9.2   Circuit synthesized from incomplete **case** statement.

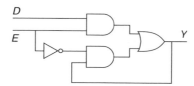

Figure 9.3   Asynchronous circuit synthesized from **when . . . else**.

type has, say, three values, the **others** clause is needed to cover the unused state(s). This is not required according to the VHDL standard, but if you really want to protect yourself against poor synthesis tools, include the **others** clause!) A similar argument applies to **with . . . select** statements in concurrent VHDL.

Note that the latch used in these examples would be taken from a library. The design of asynchronous elements such as latches and flip-flops is covered in Chapter 12. Such elements cannot be synthesized from first principles by a synthesis tool. The concurrent statement

```
y <= D when E = '1' else y;
```

in which a signal appears on both the left- and right-hand sides of the signal assignment, would be synthesized to the circuit of Figure 9.3. This is apparently functionally correct, but it contains a potential hazard and is therefore a poor latch design. Such constructs should be avoided. Conditional concurrent statements without a final **else**, or that include **unaffected** as one of the choices, should be avoided for the same reason. None of these forms is supported by the 1076.6 standard.

Edge-sensitive flip-flop

As described in Chapter 6, edge-sensitive behaviour may be modelled by putting a **wait until** statement at the top of a process:

```
p0: process is
begin
  wait until (Clock = '1');
  Q <= D;
end process p0;
```

The process suspends until Clock changes to a logic 1. Therefore any signals that have assignments made to them after the **wait** statement must hold on to their values until the **wait** statement has completed. Thus signal assignments following a **wait until** statement will be interpreted as inputs to edge-triggered flip-flops. In general, a process written using a **wait until** statement that is to be interpreted by an RTL synthesis system should have *exactly one* **wait until** statement. This **wait until** statement must be the *first executable line* of the process and must depend on only *one* transition of *one* signal. Some synthesis systems may allow more complex structures, but these restrictions are specified in IEEE 1076.6.

In Chapter 6, an alternative way of modelling flip-flops was described:

```
p0: process (Clock) is
begin
  if rising_edge(Clock) then
    Q <= D;
  end if;
end process p0;
```

The rising_edge and falling_edge functions, together with expressions such as (Clock = 1 **and** Clock'EVENT), are interpreted by a synthesis system to model edge-sensitive behaviour. Hence signal assignments that can be reached only by fulfilling an edge-sensitive condition will be interpreted as assignments to registered signals. It should be remembered that the signal name itself is not meaningful to the synthesis tool.

Asynchronous sets and resets are modelled using level-sensitive **if** clauses:

```
p0: process (Clock, Reset) is
begin
  if (Reset = '0') then
    Q <= '0';
  elsif rising_edge(Clock) then
    Q <= D;
  end if;
end process p0;
```

This structure would be interpreted, correctly, as a positive-edge triggered flip-flop with an active low asynchronous reset. The reset is tested before the clock and therefore has an effect irrespective of the clock. The clock signal to which the flip-flop is edge-sensitive should be tested in the last branch of the **if** statement. Similarly, synchronous sets and resets and clock enable inputs as described in Chapter 6 will be correctly interpreted by an RTL synthesis tool.

We saw in the previous chapter that the VHDL simulation model means that signal assignments in a process do not take effect until the subsequent time point or delta cycle. Variable assignments, on the other hand, take immediate effect. The synthesized forms of signal and variable assignments should therefore be different. The following fragment of VHDL synthesizes to the structure shown in Figure 9.4.

```
process (clock) is
begin
  if rising_edge(Clock) then
```

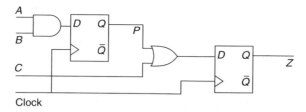

Figure 9.4 Circuit synthesized by sequential signal assignments.

```
      P <= A and B;
      Z <= P or C;
    end if;
end process;
```

In the first assignment, P, a signal, is given a value. When P is referenced in the second assignment, the new value of P has not yet taken effect. Therefore the previous value of P is used. The value of P (and of Z) is not updated until the process resumes, at the next clock edge. Therefore P behaves exactly as if its value were stored in a flip-flop.

By contrast, a variable assignment takes effect immediately. Therefore the following piece of code in which P is a variable is synthesized to the structure of Figure 9.5.

```
process (clock) is
  variable P : std_logic;
begin
  if rising_edge(Clock) then
    P := A and B;
    Z <= P or C;
  end if;
end process;
```

According to the IEEE 1076.6 standard, there should be no statements before the **if** statement or after the corresponding **end if** line in the process. The sensitivity list should consist of the clock signal followed by any other signals sensed in the asynchronous parts of the **if** statement (i.e. the asynchronous control signals and any signals assigned to other signals in the asynchronous parts). Several of the sequential

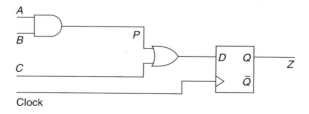

Figure 9.5 Circuit synthesized using variable assignment.

examples of Chapter 6 do not conform to this requirement, but are nevertheless acceptable to some synthesis tools.

9.1.3   Combinational logic

In general, if a piece of hardware is not a level-sensitive or edge-sensitive sequential
unit, it must be a combinational unit. Therefore, a VHDL description that does not ful-
fil the conditions for synthesis to level-sensitive or edge-sensitive sequential elements
must by default synthesize to combinational elements. Hence the problem of describing combinational hardware in VHDL is to ensure that we do not accidentally cause the
synthesis tool to infer the existence of registers.

To ensure that combinational logic is synthesized from a VHDL process, we must
observe three conditions. First, we must not have any edge-triggered statements in that
process. Secondly, if a signal or variable has a value assigned in one branch of an **if**
statement or a **case** statement, that signal or variable must have a value assigned in
every branch of the statement (or it must have a value assigned before the branching
statement – see below). Finally, all the signals sensed either as branching conditions or
in signal or variable assignments must be included in the sensitivity list of the process.

For example, the following is a model of a state machine with two states, two inputs
and two outputs.

```
library IEEE;
use IEEE.std_logic_1164.all;

entity Fsm is
  port (Clock, InA, InB : in std_logic;
        OutA, OutB : out std_logic);
end entity Fsm;

architecture Try1 of Fsm is
begin
  p0: process is
    type State is (S0, S1, S2);
    variable PresentState: State;
  begin
    wait until rising_edge(Clock);
    case PresentState is
      when S0 =>
        OutA <= '1';
        if InA = '1' then
          PresentState := S1;
        end if;
      when S1 =>
        OutA <= InB;
        OutB <= '1';
        if (InA = '1') then
```

```
        PresentState := S2;
      end if;
    when S2 =>
      OutB <= InA;
      PresentState := S0;
  end case;
  end process p0;
end architecture Try1;
```

Although this is an acceptable simulation model, if it were synthesized, OutA and OutB would be registered in addition to PresentState, because they have values assigned to them after the edge-triggered **wait until** statement. Thus we can divide the model into two processes, one combinational and one sequential.

```
architecture Try2 of Fsm is
  type State is (S0, S1, S2);
  signal PresentState, NextState: State;
begin
  s0: process is
  begin
    wait until rising_edge(Clock);
    PresentState <= NextState;
  end process s0;
  c0: process (PresentState) is
  begin
    case PresentState is
      when S0 =>
        OutA <= '1';
        if InA = '1' then
          NextState <= S1;
        else
          NextState <= S0;
        end if;
      when S1 =>
        OutA <= InB;
        OutB <= '1';
        if (InA = '1') then
          NextState <= S2;
        else
          NextState <= S1;
        end if;
      when S2 =>
        OutB <= InA;
        NextState <= S0;
    end case;
  end process c0;
end architecture Try2;
```

This will, again, simulate as a state machine giving apparently correct behaviour. When synthesized, however, OutA and OutB will be registered through asynchronous latches, because in state S0 no value is assigned to OutB and hence OutB holds on to its value. Similarly in state S2, no value is assigned to OutA.

This error can be resolved by explicitly including an assignment to both OutA and OutB in every branch of the **case** statement. Alternatively, both signals can be given default values at the start of the process:

```
c0: process (PresentState) is
begin
  OutA <= '0';
  OutB <= '0';
  case PresentState is
    when S0 =>
      OutA <= '1';
      if InA = '1' then
        NextState <= S1;
      else
        NextState <= S0;
      end if;
    when S1 =>
      OutA <= InB;
      OutB <= '1';
      if (InA = '1') then
        NextState <= S2;
      else
        NextState <= S1;
      end if;
    when S2 =>
      OutB <= InA;
      NextState <= S0;
  end case;
end process c0;
```

This process now synthesizes to purely combinational logic, while process S0 synthesizes to edge-triggered sequential logic. Most synthesis tools will (or should) give a warning, however. A piece of combinational logic will be synthesized with three inputs (PresentState, InA and InB) and three outputs (NextState, OutA and OutB). Hence a change at any of the inputs could cause a change at an output. The VHDL model above has only one signal in its sensitivity list (PresentState). Therefore this model and the synthesized circuit may behave differently when simulated. To avoid this, all the signals to which the combinational logic is sensitive should be included in the sensitivity list. The 'correct' interpretation of a model with an incomplete sensitivity list such as

```
p0: process (a) is
begin
  q <= a and b;
end process p0;
```

(a)

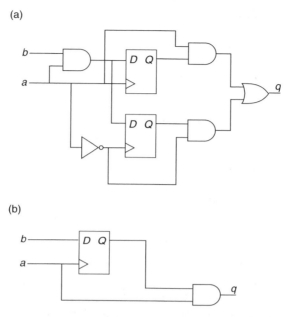

(b)

Figure 9.6 (a) Circuit synthesized from incomplete sensitivity list; (b) optimized circuit.

is the circuit shown in Figure 9.6(a). The lower flip-flop of this circuit will always have a 0 output, so in theory this circuit can be optimized to that of Figure 9.6(b).

The complete, correct model of the example state machine is shown below.

```
architecture Try3 of Fsm is
  type State is (S0, S1, S2);
  signal PresentState, NextState: State;
begin
  s0: process is
  begin
    wait until rising_edge(Clock);
    PresentState <= NextState;
  end process s0;
  c0: process (PresentState, InA, InB) is
  begin
    OutA <= '0';
    OutB <= '0';
    case PresentState is
      when S0 =>
        OutA <= '1';
        if InA = '1' then
          NextState <= S1;
```

```
        else
          NextState <= S0;
        end if;
      when S1 =>
        OutA <= InB;
        OutB <= '1';
        if (InA = '1') then
          NextState <= S2;
        else
          NextState <= S1;
        end if;
      when S2 =>
        OutB <= InA;
        NextState <= S0;
    end case;
  end process c0;
end architecture Try3;
```

The style of coding will also influence the final hardware. For example, nested **if** . . . **then** . . . **else** blocks, such as the priority encoder of Section 4.4.2, will tend to result in priority encoding and hence long chains of gates and large delays. On the other hand, **case** statements, such as the state machine above, will tend to be synthesized to parallel multiplexer-type structures with smaller delays. Similarly, shift operations will result in simpler structures than multiplication and division operators.

9.1.4   Summary of RTL synthesis rules

It is easy to make mistakes and to accidentally create latches when combinational logic is intended (or worse, to deliberately create latches, when you really want a flip-flop – see Section 5.5.4). Table 9.1 summarizes the rules for creating combinational and sequential logic from processes.

Table 9.1   Summary of RTL synthesis rules.

	Sensitivity list	Branches
Combinational logic	All inputs in sensitivity list (signals on RHS of assignments and used in **if** and **case** statements)	Complete (or default values)
Latches	All inputs in sensitivity list (signals on RHS of assignments and used in **if** and **case** statements)	Not complete
Flip-flops	Clock and asynchronous set and reset only (process usually contains **wait on** statement or `rising_edge` or `falling_edge` function)	Not complete

There is one further rule that applies to all synthesizable logic: do not assign a value to a signal in two or more processes. The only exception to this rule is the case of three-state logic, as in the bus in the microprocessor example of Chapter 7. You should be able to draw a block diagram of your design, with each process represented by a box. If two boxes appear to be driving the same wire, you have done something wrong. (Indeed, if you can't draw the block diagram, you have made a really serious mistake!)

9.2 Constraints

For any non-trivial digital function, there exist a number of alternative implementations. Ideally, a digital system should be infinitely fast, infinitesimally small, consume no power and be totally testable. In reality, of course, this ideal is impossible. Therefore, the designer has to decide what his or her objectives are. These objectives are expressed to the synthesis tool as *constraints*. Typically, a design has to fit on a particular FPGA and has to operate at a particular clock frequency. Thus two constraints of area and speed have to be specified. It is possible that these constraints will be in conflict. For example, a design may fit on a particular FPGA, but not work at the desired speed – to reach the desired speed may require more logic and hence more area, as illustrated in Figure 9.7. Assuming that CMOS logic is used and that the gate delays are identical, the circuit of Figure 9.7(a) would need 16 transistors and have a maximum delay of 4 units, while the circuit of Figure 9.7(b) requires 18 transistors and has a maximum delay of 3 units.

9.2.1 User-defined attributes

Synthesis constraints can be expressed in two ways: as VHDL **attributes** in the model description or as some other format in a separate file. There is no standard between tools for either the type of constraints or the format in which they may be expressed. In general, user-defined attributes are used to pass information to synthesis tools, but are ignored by simulators.

An attribute definition has two parts. In the first part, the name of the attribute and its type are declared. In the second part, the attribute is associated with a VHDL entity and given a value. In this context 'entity' can refer to an **entity, architecture,**

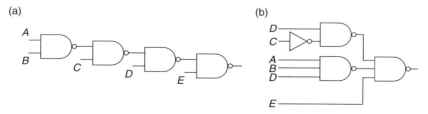

Figure 9.7 Two versions of a combinational circuit: (a) smaller, slower; (b) larger, faster.

configuration, **procedure**, **function**, **package**, **type**, **subtype**, **constant**, **signal**, **variable**, **component**, **label**, **literal**, **units**, **group** or **file**.

For example, IEEE 1076.6 defines one attribute for specifying the state encoding, e.g.

```
attribute enum_encoding : STRING;
type state is (s0, s1, s2, s3);
attribute enum_encoding of state: type is "00 01 11 10";
```

This might instead be expressed in a separate constraints file using a format like:

```
enum_encoding : state = '00 01 11 10'
```

Other example attribute definitions could be as follows:

```
attribute pin_no : NATURAL;
attribute pin_no of q : signal is 42;

attribute technology : STRING;
attribute technology of all : component is "CMOS";
```

Given that the type and format of constraints are unique to particular synthesis tools, in the following sections we will discuss only the general types of constraints that can be specified.

9.2.2   Area and structural constraints

State encoding

As discussed in Chapter 5, a state machine with s states can be implemented using m state variables, where

$$2^{m-1} < s \leq 2^m$$

There are

$$\frac{(2^m)!}{(2^m-s)!}$$

possible state assignments. There is no method for determining which of these assignments will result in minimal combinational next state logic. In addition, other non-minimal state encoding schemes, such as one-hot, exist. No RTL synthesis tools attempt to tackle the general state assignment problem. Heuristic methods may be able to choose either a binary counting sequence or one-hot encoding. Therefore one design constraint that can be specified is the state encoding method, either using the IEEE 1076.6 style or by specifying the code with a keyword, as shown above.

Resource constraints

The use of a particular technology may constrain the type of structures that can be created. Features of different FPGA technologies will be discussed later in this chapter. Having selected a particular technology, a range of different-sized devices may exist, and very often it is desirable to select the smallest possible. Thus the specification of a particular device is a constraint on the synthesis process.

As a single ASIC or FPGA has to be connected via a printed circuit board to other devices, the functionality of each pin may have to be determined in advance of the synthesis. Therefore another constraint is the association of a signal with a particular pin.

Under some circumstances, complex logic blocks may be reused. For example, the following piece of code can be implemented with two adders or with one adder and two multiplexers:

```
if Select = '1' then
  q <= a + b;
else
  q <= c + d;
end if;
```

A synthesis constraint can choose whether resources are shared, either at a local level or globally. Such choices have implications for both the area and speed of the final design.

Finally, it may be desirable to describe a function in VHDL in order to verify the correct operation of the rest of the system, but when the system is synthesized we would rather use a predefined library component to implement that function instead of synthesizing the function from first principles. Therefore we can designate that a particular unit is a 'black box' that we will incorporate from a library, e.g.

```
attribute black_box : BOOLEAN;
attribute black_box of b0 : component is TRUE;
```

Timing constraints

If we want a circuit to operate synchronously with a clock at a particular frequency, say 20 MHz, we know that the maximum delay through the state registers and the next state logic is the reciprocal of the clock frequency, in this case 50 ns. Therefore a constraint on the synthesis tool can be expressed as the clock frequency or as the maximum delay through the combinational logic, as shown in Figure 9.8.

The difficulty, from the synthesis point of view, with this approach is that the delay through the combinational logic can only be estimated. The exact delay depends on how the combinational logic is laid out, and hence the delay depends on the delay through the interconnect. Therefore the synthesis is performed using an estimate of the likely delays. Having generated a netlist, the low-level place and route tool attempts to fit the design onto the ASIC or FPGA. The place and route tool can take into account the design constraint – the maximum allowed delay – and the delays through the logic that has been generated. At this stage, it may become apparent that the design objective

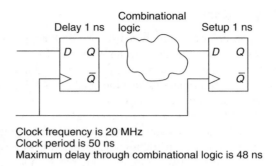

Clock frequency is 20 MHz
Clock period is 50 ns
Maximum delay through combinational logic is 48 ns

Figure 9.8   Basic timing constraint.

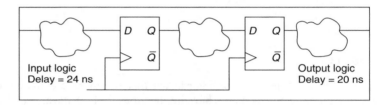

Figure 9.9   Input and output timing constraints.

cannot be achieved, so the design would have to be synthesized again with a tighter
timing constraint to allow for the extra time in the routing. This can mean that the final
goal is never reached. To speed up hardware more operations are performed concurrently, which means that the design is larger. Hence the design is harder to place and
route, and hence the routing delays increase, *ad infinitum*.

    More specific timing constraints can be applied to selected paths. If a design is split
between two or more designers, the signal path between registers in two parts of the
design may include combinational logic belonging to both parts of the design. If both
parts of combinational logic were each synthesized without allowing for the existence
of the other, the total delay between registers could be greater than one clock period.
Therefore timing constraints can be placed upon paths through the input and output
combinational logic in a design, as shown in Figure 9.9.

9.3   Synthesis for FPGAs

In principle, an RTL model of a piece of hardware coded in VHDL can be synthesized
to any target technology. In practice, the different technologies and structures of ASICs
and FPGAs mean that certain constructs will be more efficiently synthesized than
others and that some rewriting of VHDL may be needed to attain the optimal use of a
particular technology.

In this section we will compare two FPGA technologies and show how the VHDL coding of a design can affect its implementation in a technology. The descriptions of the technologies are deliberately simplified.

Xilinx FPGAs are based on static RAM technology. Each FPGA consists of an array of the configurable logic blocks (CLBs) shown in Figure 1.15. Each logic block has two flip-flops and a combinational block with eight inputs. Each flip-flop has an asynchronous set and reset, but only one of these may be used at one time. Each flip-flop also has a clock input that can be positive or negative edge-sensitive, and each flip-flop has a clock enable input. In addition to the CLB shown, a number of three-state buffers exist in the array.

Actel FPGAs are based on antifuse technology. Two types of logic block exist in more or less equal numbers – a combinational block and a sequential block as shown in Figure 1.14. Each flip-flop in a sequential block has an asynchronous reset.

Both types of FPGA therefore have a relatively high ratio of flip-flops to combinational logic. Conventional logic design methods tend to assume that flip-flops are relatively expensive and combinational logic is relatively cheap, and that therefore sequential systems such as state machines should be designed with a minimal number of flip-flops. The large number of flip-flops in an FPGA and the fact that the flip-flops in a Xilinx FPGA or in an Actel sequential block cannot be used without the combinational logic reverses that philosophy and suggests that one-hot encoding is a more efficient state encoding method, particularly for small state machines.

Similarly, a single global asynchronous set or reset is the most efficient way of initializing both types of FPGA. If both set and reset are required it is necessary to use additional combinational logic, hence it is better to have, for example, an asynchronous reset and a *synchronous* set.

In both technologies, the flip-flops are edge-sensitive; therefore level-sensitive latches have to be synthesized from combinational logic. Again, this can waste flip-flops, so level-sensitive designs are best avoided. It is, however, reasonable to assume that any level-sensitive latches will exist as library elements and therefore that they will be hazard-free.

In both technologies, it may be desirable to instantiate predefined library components for certain functions. Not only is the logic defined, but the configuration of logic blocks is already known, potentially simplifying both the RTL synthesis and place and route tasks.

All the foregoing comments distinguish synthesis to FPGAs from synthesis to ASICs in general. The FPGA technologies themselves favour certain VHDL coding styles. For example, the following piece of VHDL shows two ways of describing a 5-to-1 multiplexer.

```vhdl
library IEEE;
use IEEE.std_logic_1164.all;

entity Mux is
  port (a, b, c, d, e: in std_logic;
        s: in std_logic_vector(4 downto 0);
        y: out std_logic);
end entity Mux;
```

```
architecture Version1 of Mux is
begin
  p0: process (s, a, b, c, d, e) is
  begin
    case s is
      when "00001" => y <= a;
      when "00010" => y <= b;
      when "00100" => y <= c;
      when "01000" => y <= d;
      when others => y <= e;
    end case;
  end process p0;
end architecture Version1;

architecture Version2 of Mux is
begin
  y <= a when s(0) = '1' else 'Z';
  y <= b when s(1) = '1' else 'Z';
  y <= c when s(2) = '1' else 'Z';
  y <= d when s(3) = '1' else 'Z';
  y <= e when s(4) = '1' else 'Z';
end architecture Version2;
```

These two models have the same functionality when simulated. If version 1 were synthesized to a Xilinx FPGA, two CLBs would be needed. Version 2, on the other hand, can be implemented using the three-state buffers that exist outside the CLBs. Version 2, however, cannot be synthesized to an Actel FPGA as the technology does not support three-state logic, except at the periphery of the FPGA. Clearly, therefore the choice of architecture depends upon which technology is being used.

The two technologies have different limitations with respect to fan-outs. Antifuse technology has a fan-out limit of about 16 (one output can drive up to 16 inputs without degradation of the signal). CMOS SRAM technology has a higher fan-out limit. In practice, this means that a design that can easily be synthesized to a Xilinx FPGA cannot be synthesized to an Actel FPGA without rewriting. For example, an apparently simple structure such as the following fragment cannot be synthesized as it stands because the Enable signal is controlling 32 multiplexers:

```
signal a, b : std_logic_vector(31 downto 0);
begin
  p0 : process (Enable, b) is
  begin
    if Enable = '1' then
      a <= b;
    else
      a <= (others => '0');
    end if;
  end process p0;
end;
```

Instead, the `Enable` signal must be split into two using buffers, and each buffered signal then controls half the bus:

```
signal a, b : std_logic_vector(31 downto 0);
signal En0, En1 : std_logic;
begin
  b0 : buf port map (Enable, En0);
  b1 : buf port map (Enable, En1);
  p0 : process (En0, En1, b) is
  begin
    if En0 = '1' then
      a(15 downto 0) <= b(15 downto 0);
    else
      a(15 downto 0) <= (others => '0');
    end if;
    if En1 = '1' then
      a(31 downto 16) <= b(31 downto 16);
    else
      a(31 downto 16) <= (others => '0');
    end if;
  end process p0;
end;
```

A good synthesis tool should recognize the fan-out limits and automatically insert buffers.

9.4 Behavioural synthesis

In RTL synthesis, the design is specified in terms of register operations and transformed automatically into gates and flip-flops. Behavioural synthesis takes the process one stage further. The hardware to be synthesized is described in terms of an algorithm, from which the registers and logic are derived. In principle, it is not necessary to use a hardware description language for behavioural synthesis; indeed, subsets of conventional programming languages such as C have been used. The major obstacle to the widespread acceptance of behavioural synthesis appears to be the difficulty that a hardware designer has in interpreting the output of a synthesis tool. The output of RTL synthesis, particularly when expressed in terms of FPGA netlists, can be very difficult to interpret. This is even truer of behavioural synthesis, where the detailed structure is entirely generated by the synthesis tool. With the decreasing cost of silicon, however, it seems safe to predict that behavioural synthesis will become an accepted design tool, in the same way that compilers for high-level programming languages are now accepted, even though the machine code generated is largely unintelligible.

This section will show, by example, how a behavioural synthesis tool might generate a structural representation of a circuit from a high-level algorithmic description.

The following is a behavioural model of an infinite impulse response (IIR) filter.

```vhdl
package iir_defs is
  constant precision: POSITIVE := 16;
  subtype int is INTEGER range - 2**(precision - 1) to
                                 2**(precision - 1) - 1;
  type integer_array is array (NATURAL range <>) of int;
  constant order: POSITIVE := 5;
end package iir_defs;

use work.iir_defs.all;
entity iir is
  generic (coeffa: integer_array (0 to order);
           coeffb: integer_array (0 to order - 1));
  port(input: in int; strobe: in BIT; output: out int);
end entity iir;

architecture behaviour of iir is
begin
  process is
    variable input_sum: int;
    variable output_sum: int;
    variable delay: integer_array (0 to order) :=
                    (others => 0);
  begin
    input_sum := input;
    for j in 0 to order - 1 loop
      input_sum := input_sum + (delay(j)*coeffb(j))/1024;
    end loop;
    output_sum := (input_sum*coeffa(order))/1024;
    for k in 0 to order loop
      output_sum := output_sum +
                    (delay(k)*coeffa(k))/1024;
    end loop;
    for l in 0 to order - 1 loop
      delay(l) := delay (l + 1);
    end loop;
    delay(order) := input_sum;
    output <= output_sum;
    wait on strobe;
  end process;
end architecture behaviour;
```

This is a behavioural description in the sense that the filter is described purely as an algorithm. A C version of the algorithm would look very similar. A C version might not include the strobe signal, but conversely this is not RTL VHDL, as there is neither a clock nor a reset. If this description were used for RTL synthesis (assuming the synthesis tool accepted the VHDL), the resulting hardware would have 12 16-bit combinational multipliers and 11 16-bit adders. This translates to 442 544 full adders, requiring several tens of FPGAs! The division by 1024 is simply a scaling operation and can be achieved by throwing away the 10 least significant bits from each multiplication product. This operation is therefore effectively free.

The essential fact about behavioural synthesis is that it is possible to make design decisions and to achieve a compromise between speed and size. In the IIR example above, it would be equally possible to implement the algorithm using 442 544 full adders and complete the operation in one clock cycle, or to use one full adder and take 442 544 clock cycles to achieve the result. More sensibly, some implementation between these two extremes might be sufficiently fast and sufficiently small to satisfy the requirements of the final application.

It is not practical to demonstrate the principles of behavioural synthesis with the fifth-order IIR filter. Instead, let us consider how a first-order filter might be built. In order to know which operations can be done concurrently and which require successive clock cycles, we need to know the dependency of each piece of data on each other piece of data. To do this, the loops in the behavioural description will first be expanded. We will ignore the division operations for the reason stated above.

```
input_sum  := input + delay(0)*coeffb(0);
output_sum := input_sum*coeffa(1);
output_sum := output_sum + delay(0)*coeffa(0);
output     := output_sum + delay(1)*coeffa(1);
```

Assignments are made to `output_sum` on the second and third lines. To distinguish between successive values of `output_sum`, the two values will be separated, such that there is only one assignment to each variable in the algorithm. This is known as *single assignment form*.

```
input_sum   := input + delay(0)*coeffb(0);
output_sum0 := input_sum*coeffa(1);
output_sum1 := output_sum0 + delay(0)*coeffa(0);
output      := output_sum1 + delay(1)*coeffa(1);
```

From this a data dependency graph can be constructed (Figure 9.10).

If the operations shown in Figure 9.10 were all performed in one clock cycle, three adders and four multipliers would be needed. If it were decided, however, that each multiplication and each addition takes one clock cycle, the data dependency graph can be used to construct a *schedule* that shows when each operation can be performed (Figure 9.11).

It can be seen that five clock cycles are required to perform the arithmetic operations – the system is said to have a *latency* of five. This schedule is known as an *as soon as possible* (ASAP) schedule, because each operation is done as early as possible. Note that the sequence of operations is not the same as given by the original VHDL description. Equally, it is possible to schedule operations *as late as possible*

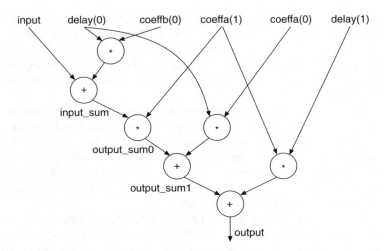

Figure 9.10 Data dependency graph.

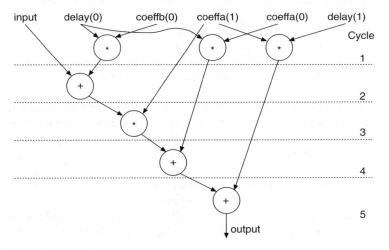

Figure 9.11 ASAP schedule.

(ALAP) (Figure 9.12). This schedule also takes five clock cycles. If, however, the resources were constrained to a single arithmetic unit, again using an ALAP schedule, the number of cycles required increases (Figure 9.13).

Given the assumption that the basic resources available are arithmetic units, there are relatively few possible schedules for this example. With larger problems, the number of possible schedules clearly increases. By limiting the available resources, and hence the total area of the design, the latency, i.e. the time taken to complete an operation, is

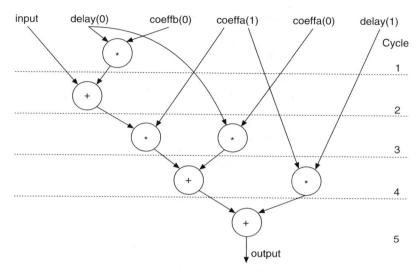

Figure 9.12   Unconstrained ALAP schedule.

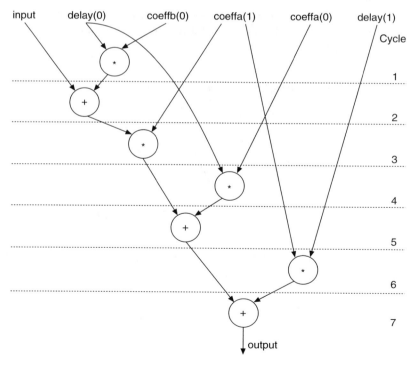

Figure 9.13   Resource constrained schedule.

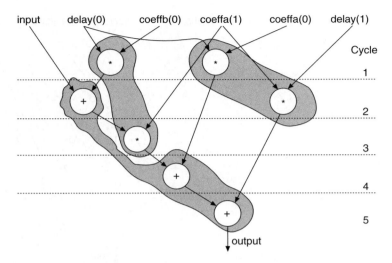

Figure 9.14 Mapping of operations onto resources.

increased. Therefore the synthesis tool can trade speed against area by changing the schedule.

Figure 9.14 shows how the operations can be mapped onto particular resources. The three shaded groups each represent a resource used in different clock cycles, namely two multipliers and an adder.

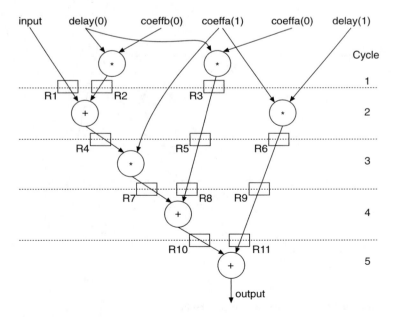

Figure 9.15 Schedule showing registers.

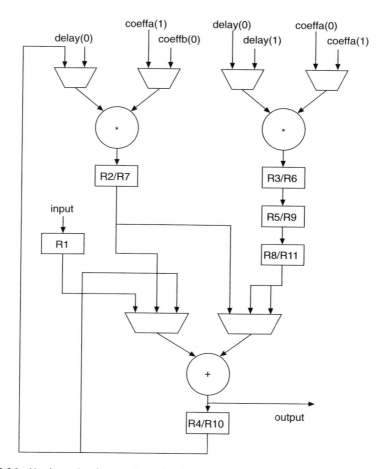

Figure 9.16 Hardware implementation of first-order filter.

The result of an operation is used in a subsequent clock cycle. Therefore every time a data arc crosses a clock boundary a register must be inserted, as shown in Figure 9.15.

Just as the arithmetic resources can be shared, so too can the registers be shared. The sharing is achieved using multiplexers, which are assumed to be cheap (i.e. small) compared with the other resources. Hence a possible hardware implementation of the schedule of Figure 9.15 is shown in Figure 9.16.

In Figures 9.15 and 9.16, three registers are shown following one of the multiplier units. This assumes that every register is loaded at each clock edge. It would be equally valid to use enabled registers, which would reduce the number of registers. Whatever technique is used, the multiplexers and registers have to be controlled. We have so far discussed the derivation of the datapath part of Figure 7.6 from a behavioural description. The controller part also needs to be synthesized. In the example shown this is relatively simple. There are five clock cycles, hence five states as shown in Figure 9.17.

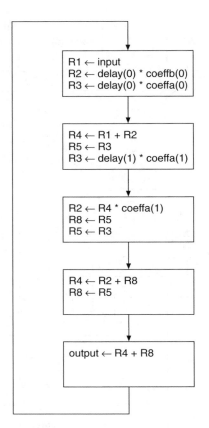

Figure 9.17   ASM chart of controller.

9.5 Verifying synthesis results

Synthesis should, by definition, produce a correct low-level implementation of a design from a more abstract description. In principle, therefore, functional verification of a design after synthesis should not be needed. For peace of mind, we might wish to check that the synthesized design really does perform the same function as the RTL description. Synthesis does, however, introduce an important extra factor to a design – timing. An RTL design is effectively cycle-based. A task takes a certain number of clock cycles to complete, but we do not really know how long each cycle takes. After synthesis, the design is realized in terms of gates or other functional blocks, and these can be modelled with delays. After placement and routing, we have further timing information in the form of wiring delays, which can be significant and which can affect the speed at which a design can operate.

It is possible, in principle, to verify a synthesized design by comparing it with the original RTL design, using techniques such as model-checking. In practice, such tools are limited to checking interfaces. Static timing analysis can give us information about delays between two points in a circuit, but cannot distinguish between realizable signal paths and false paths that are never enabled in reality. Similarly, a synthesis tool aims to meet timing constraints, but cannot distinguish between true and false paths. Therefore the only way to verify the timed behaviour of a synthesized system is to simulate it.

One approach to checking a design at two levels of abstraction is to simulate both versions at the same time and to compare the results. This is usually a bad idea for two reasons. First, the size of the system to be simulated is at least twice as large as one version in isolation, and therefore slower to execute. Second, there will, as noted, be timing differences. Therefore comparing responses may lead to false warnings.

The testbench design examples described in Chapters 3, 4, 5 and 6 are well suited to simulating post-synthesis designs. In particular, the idea of checking a response by synchronizing to the clock and then waiting for the signal to stabilize is very appropriate for checking timing responses.

9.5.1   VITAL and SDF

VITAL (VHDL Initiative Towards ASIC Libraries, 1076.4-2000) is a set of low-level VHDL primitives for accurate timing simulations of gate-level models and wires. These primitives can be used to construct models of FPGA or ASIC cells. There are two advantages to using cell libraries built from VITAL components. First, the standard models can be *accelerated* in a simulator. In other words, instead of compiling and simulating VHDL models, these models can be built into the simulator, giving much better simulation speed than arbitrary VHDL models. Second, the simulator can associate Standard Delay Format (SDF) files with the VITAL models. SDF is a standard (IEEE 1497-2001) for describing delays in netlist files. It is not part of VHDL – SDF may be used with any type of netlist file.

After both RTL synthesis and place and route, the VITAL-compliant gate and wire models and the SDF file for a design can be extracted. An SDF file typically has minimum, typical and maximum delays for each piece of logic. When the VITAL netlist file is compiled and loaded into the simulator, one set of values can also be loaded. Thus, timing simulations can be performed that accurately reflect the behaviour of the real circuit can be performed.

We will not describe the format of VITAL files here. (There is a brief example of a VITAL file in Chapter 12.) In general, it is extremely difficult to interpret an automatically generated netlist file. Cell and wire names are usually obscure and all but the simplest designs are too complex to understand. Moreover, it is very unlikely that you would ever need to write a VITAL netlist or to build VITAL-compliant models. It is sufficient to know that the extracted netlist will have an entity description similar to the RTL model that was originally synthesized. Be warned, however, that the extracted entity description could have signal ports declared in a different order from your original RTL design, that there are unlikely to be any **generic**s, and that types such as `integer` or `signed` will have been converted to `std_logic_vector`. Therefore, your original testbench may need some modification. You would normally treat the architecture part of the netlist as a black box and simply observe output signals.

Summary

VHDL was conceived as a description language, but has been widely adopted as a specification language for automatic hardware synthesis. A number of tools exist for RTL synthesis, but behavioural synthesis tools are appearing. Because of its origins, VHDL has some features that are not synthesizable to hardware. The rules for the inference of latches and flip-flops are well defined. Synthesis constraints may be stated in terms of VHDL attributes or as separate inputs to the synthesis tool. To get the most out of an FPGA may require careful writing of the VHDL code. The important concepts behind behavioural synthesis are scheduling and binding.

Further reading

Despite the definition of a synthesizable subset of VHDL, each synthesis tool accepts a slightly different subset of VHDL and interprets poorly written VHDL in different ways. It therefore pays to read the user manuals of tools with some care. The websites of FPGA manufacturers include VHDL style guides showing what can and cannot be implemented.

De Micheli covers both high-level behavioural synthesis and low-level optimization in his book.

Exercises

9.1 Explain, with examples, what is meant by a *constraint* in RTL synthesis.

9.2 Write a model of an eight-state counter as a VHDL *state machine*, with a clock and reset inputs, which outputs a ready flag when the counter is in the initial state. Use the `enum_encoding` attribute to specify that the state machine should be implemented as a Johnson counter.

9.3 The listing below shows a description of a simple state machine in VHDL. If this state machine were synthesized using an RTL synthesis tool, the resulting hardware would give different simulated behaviour from the original RTL description. Explain why this should be so.

```
library ieee;
use ieee.std_logic_1164.all;

entity fsm is
  port (clk, a: in std_logic;
        y: out std_logic);
end entity fsm;

architecture try1 of fsm is
  type statetype is (s0, s1, s2);
```

```
    signal currentstate, nextstate : statetype := s0;
begin
  seq: process (clock) is
  begin
    if rising_edge(clock) then
      currentstate <= nextstate;
    end if;
  end process seq;
  com: process (currentstate) is
  begin
    case currentstate is
      when s0 =>
        if a = '1' then
          nextstate <= s1;
        else
          nextstate <= s2;
        end if;
      when s1 =>
        y <= '1';
        nextstate <= s0;
      when s2 =>
        if a = '1' then
          nextstate <= s2;
        else
          nextstate <= s0;
        end if;
    end case;
  end process com;
end architecture try1;
```

9.4 Rewrite the VHDL model of Exercise 9.3 such that, when synthesized, the resulting hardware consists only of D flip-flops, with asynchronous resets and combinational next state and output logic.

9.5 The listing below shows three VHDL processes. Describe the hardware that
should be generated from each of these process models by a synthesis tool.

```
architecture abc of abc is
begin
  a: process (x, y) is
  begin
    if y = '1' then
      qa <= x;
    else
      qa <= '0';
    end if;
  end process a;
```

```
  b: process (x, y) is
  begin
    if y = '1' then
      qb <= x;
    end if;
  end process b;
  c: process (y) is
  begin
    if y = '1' then
      qc <= x;
    end if;
  end process c;
end architecture abc;
```

9.6 Explain the terms *scheduling* and *binding* in the context of behavioural synthesis.

9.7 The following sequence of operations is part of a cube root solution routine:

```
a <= x * x;
a <= 3 * a;
b <= y / a;
a <= 2 * x;
a <= a / 3;
c <= a - b;
```

Convert this sequence to *single assignment form* and hence construct a data dependency graph. Assuming that each arithmetic operation takes exactly one clock cycle, derive an *unconstrained as late as possible (ALAP)* schedule.

9.8 Derive a *constrained* schedule for the routine of Exercise 9.7 and hence design a datapath implementation of this part of the system, assuming that one multiplier, one divider and one subtracter are available.

Chapter 10

Testing digital systems

In the course of manufacture, defects may be introduced into electronic systems. Systems may also break during use. Defects may not be easy to detect. In this chapter we will discuss the importance of testing, the types of defect that can occur and how defects can be detected. We describe procedures for generating tests and how the effectiveness of tests can be assessed. We conclude with a technique for performing fault simulation in VHDL.

10.1 The need for testing

No manufacturing process can ever be perfect. Thus, real electronic systems may have manufacturing defects such as short circuits, missing components or damaged components. A manufacturer needs to know whether a system (whether at the level of a board, an IC or a whole system) has a defect and therefore does not work in some way. While a manufacturer does not want to sell bad systems, equally he or she would not want to reject good systems. Therefore the need for testing is economic.

We also need to distinguish between the ideas of *verification*, in which the design of a piece of hardware or software is checked, and of *testing*, in which it is assumed that the design is correct but that there may be manufacturing faults. This chapter is about the latter concept, but the inclusion of design for test structures *may* help in verifying and debugging a design.

There are, in general, two approaches to testing. We can ask whether the system works correctly (*functional* testing) or we can ask whether the system contains a fault (*structural* testing). These two approaches might at first appear to be equivalent, but in fact the tactic we adopt can make a profound difference to how we develop tests and how long those tests take to apply. Functional testing can imply a long and difficult task because all possible states of a system have to be checked. Structural testing is often easier, but is dependent upon the exact implementation of a system.

10.2 Fault models

An electronic system might contain a large number of possible defects as a result of the manufacturing process. For example, the printed circuit board could have breaks in connections because of bad etching, stress or bad solder joints. Equally there may be short circuits resulting from the flow of solder. The components on a PCB may be at fault – so-called 'population defects' – caused by having the wrong components, wrongly inserted components or omitted components. The components themselves may fail because the operating conditions exceed the component specifications or because of electromagnetic interference (EMI) or heat.

Similar defects can occur in integrated circuits. Open circuits may arise from electromigration (movement of metal atoms in electromagnetic fields), current overstress or corrosion. Silicon or oxide defects, mask misalignment, impurities and gamma radiation can cause short circuits and incorrect transistor operation. 'Latch-up', caused by transient currents, forces the output of a CMOS gate to be stuck at a logic value. In memory circuits there may be data corruption because of alpha-particles or EMI.

Clearly, to enumerate and check for every possible defect in an electronic system would be an enormous task. Therefore a distinction is made between physical *defects* and electrical *faults*. The principle of fault modelling is to reduce the number of effects to be tested by considering how defects manifest themselves. A physical defect will manifest itself as a logical fault. This fault may be static (e.g. shorts, breaks), dynamic (components out of specification, timing failures) or intermittent (environmental factors).

The relative probabilities of faults that appear during tests in manufacturing are shown in Figure 10.1. Dynamic faults may be further divided into timing faults (28%)

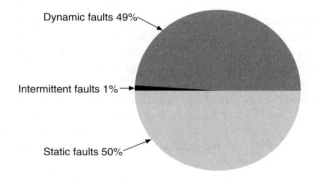

Figure 10.1 Fault probabilities.

and driver faults (21%). Timing faults and intermittent faults may be due to poor design. It is difficult to design test strategies for such faults.

10.2.1   Single-stuck fault model

Static faults are usually modelled by the *stuck fault model*. Many physical defects can be modelled as a circuit node being either stuck at 1 (s-a-1) or stuck at 0 (s-a-0). Other fault models include stuck open and stuck short faults. Programmable logic and memory have other fault models.

The *single-stuck fault model* (SSFM) assumes that a fault directly affects only one node and that the node is stuck at either 0 or 1. These assumptions make test pattern generation easier, but the validity of the model is questionable. Multiple faults do occur and multiple faults can theoretically mask each other. On the other hand, the model appears to be valid most of the time. Hence, almost all test pattern generation relies on this model. Multiple faults are generally found with test patterns for single faults.

10.2.2   PLA faults

PLAs consist not of gates, but of AND and OR logic planes, connected by fuses (or antifuses). Thus faults are likely to consist of added or missing fuses, not stuck faults. For example, Figure 10.2 shows part of a PLA where the output Z is the logical OR of three intermediate terms, P, Q and R. Each of the intermediate terms is the AND of the three inputs, A, B and C, or its inverse:

$$Z = P + Q + R$$
$$P = B.\overline{C}$$
$$Q = A.C$$

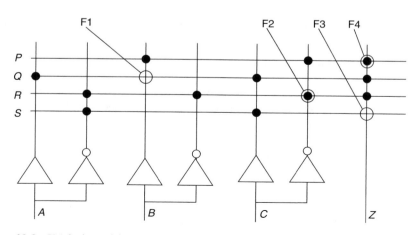

Figure 10.2   PLA fault models.

$$R = \overline{A}.\overline{B}.\overline{C}$$
$$S = \overline{A}.C$$

Fault F1 is an additional connection causing Q to change from $A.C$ to $A.B.C$. On a Karnaugh map this represents a decrease in the number of 1s circled; therefore this can be thought of as a *shrinkage* fault.

Fault F2 is a missing connection, causing R to *grow* from $\overline{A}.\overline{B}.\overline{C}$ to $\overline{A}.\overline{B}$.

Fault F3 causes the *appearance* of term S in Z.

Fault F4 causes the *disappearance* of term P from Z.

(10.3) Fault-oriented test pattern generation

Having decided that defects in a system can be modelled as electrical faults, we then need to determine whether or not any of these faults exist in a particular instance of a manufactured circuit. If the circuit were built from discrete transistors or gates, this task could, in theory, be achieved by monitoring the state of every node of the circuit. If the system is implemented as a packaged integrated circuit, this approach is not practical. We can observe only the outputs of the system and we can control only the inputs of the system. Therefore the task of test pattern generation is that of determining a set of inputs to indicate unambiguously whether an internal node is faulty. If we consider only combinational circuits for the moment, the number of possible input combinations for an n-input circuit is 2^n. We could apply all 2^n inputs (in other words, perform an exhaustive functional test), but in general we want to find the minimum necessary number of input patterns. It is possible that, because of the circuit structure, certain faults cannot be detected. Therefore it is common to talk about the *testability* of a circuit.

Testability can be a somewhat abstract concept. One useful definition of testability breaks the problem into two parts:

- *Controllability* – can we control all the nodes to establish whether there is a fault?
- *Observability* – can we observe and distinguish between the behaviour of a faulty node and that of a fault-free node?

In order to generate a minimum number of test patterns, a fault-oriented test generation strategy is adopted. In the pseudocode below, a *test* is one set of inputs to a (combinational) circuit. The overall strategy is as follows.

- Prepare a fault list (e.g. all nodes stuck-at-0 and stuck-at-1).
- Repeat:
  – write a test
  – check fault cover (one test may cover more than one fault)
  – (delete covered faults from list)
- until fault cover target is reached.

Test pattern generation (writing a test) may be random or optimized. This will be discussed in more detail below. One test may cover more than one fault; often faults are indistinguishable. Again this is discussed later.

If we simply want a pass/fail test, once we have found a test for a fault, we can remove faults from further consideration. If we want to diagnose a fault (for subsequent repair) we probably want to find all tests for a fault to deduce where the fault occurs. The fault cover target may be less than 100%. For large circuits, the time taken to find all possible tests may be excessive. Moreover, the higher the cover, the greater the number of tests and hence the cost of applying the test.

10.3.1   Sensitive path algorithm

The circuit of Figure 10.3 has seven nodes, therefore there are 14 stuck faults:

$$A/0, A/1, B/0, B/1, C/0, C/1, D/0, D/1, E/0, E/1, F/0, F/1, Z/0, Z/1$$

where $A/0$ means 'A stuck-at-0', etc.

To test for $A/0$, we need to set A to 1 (the fault-free condition – if A were at 0, we would not be able to distinguish the faulty condition from the fault-free state). The presence or otherwise of this fault can be detected only by observing node Z. We now have to determine the states of the other nodes of the circuit that allow the state of A to be deduced from the state of Z. Thus we must establish a sensitive path from A to Z. If node B is 0, E is 1 irrespective of the state of A. Therefore, B must be set to a logical 1. Similarly if F is 1, Z is 1, irrespective of E; hence F must be 0. To force F to 0, either C or D or both must be 0.

Thus, if the fault $A/0$ exists, E is 1 and Z is 1. If it does not exist, E is 0 and Z is 0.

We can conclude from this that a test for $A/0$ is $A = 1, B = 1, C = 0, D = 1$, for which the fault-free output is $Z = 0$. This can be expressed as 1101/0. Other tests for $A/0$ are 1110/0 and 1100/0. Therefore, there is more than one test for the fault $A/0$.

Let us now consider a test for another fault. To test for $E/1$ requires that $F = 0$ to make E visible at Z. Therefore C or D or both must be 0. To make $E = 0$ requires that $A = B = 1$. So a test for $E/1$ is 1101/0. This is the same test as for $A/0$. So one test can cover more than one fault.

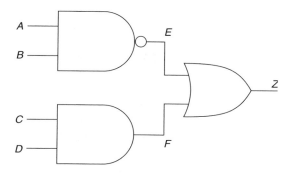

Figure 10.3   Example circuit for test generation.

The sensitive path algorithm therefore consists of the following steps:

1. Select a fault.
2. Set up the inputs to force the node to a fixed value.
3. Set up the inputs to transmit the node value to an output.
4. Check that the input node values for steps 2 and 3 are consistent.
5. Check for coverage of other faults.

The aim is to find the minimum number of tests that cover all the possible faults, although 100% fault cover may not be possible.

Fan-out and reconvergence can cause difficulties for this algorithm. Improved algorithms (D-algorithm, PODEM) use similar techniques, but overcome these drawbacks.

10.3.2   Undetectable faults

Consider the function

$$Z = A.C + B.\overline{C}$$

To avoid hazards, the redundant term may be included, as shown in Figure 10.4:

$$Z = A.C + B.\overline{C} + A.B$$

We will now try to find a test for $F/0$. This requires that F be set to 1. Hence, $A = B = 1$. To transmit the value of F to Z means that $D = E = 0$ (otherwise Z would be 1, irrespective of F). For E to be 0, B must be 0 and/or C must be 1. Similarly, for D to be 0, A must be 0 and/or C must be 0. These three conditions are inconsistent, so no test can be derived for the fault $F/0$.

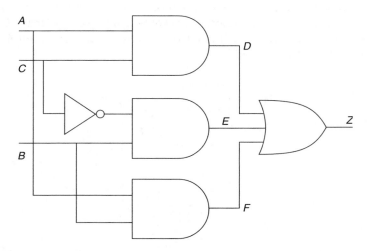

Figure 10.4   Circuit with redundancy.

There are three possible responses to this. Either it must be accepted that the circuit is not 100% testable; or the redundant gate must be removed, risking a hazard; or the circuit must be modified to provide a control input for testing purposes, to force D to 0 when $A = C = 1$.

In general, untestable faults are due to redundancy. Conversely, redundancy in combinational circuits will mean that those circuits are not fully testable.

10.3.3   The D-algorithm

The simple sensitized path procedure does not handle reconvergent paths adequately. For example, consider the circuit of Figure 10.5. To find a test for $B/1$ requires that B be set to 0. To propagate the state of B to E requires that A is 1, and to propagate E to Z requires that F is 0. The only way that F can be at 0 is if B and C are both 1, but this is not the case when $B = 0$. Apparently, therefore, the sensitive path algorithm cannot find a test for $B/1$. In fact, 101/1 is a suitable test, because under fault-free conditions E, F and Z are all at logical 1; when $B/1$, all three nodes are at logical 0.

The D-algorithm overcomes this problem by introducing a five-valued algebra: $\{0, 1, D, \overline{D}, X\}$. D represents a node that is logical 1 under fault-free (normal) conditions and logical 0 under faulty conditions. $\overline{D}$ represents a normal 0, and a faulty 1. X is an unknown value. The values of D and $\overline{D}$ are used to represent the state of a node where there is a fault and also the state of any other nodes affected by the fault.

The D-algorithm works in the same way as the sensitive path algorithm, above. If step 4 fails, the algorithm backtracks. In both steps 2 and 3 it is possible that more than one combination of inputs generates the required node values. If necessary, all possible combinations of inputs are examined.

Table 10.1 shows the inputs required to establish a fault at an internal node, to transmit that fault to an output, and to generate a fixed value (to establish or propagate a fault). Finally, it shows how fault conditions can reconverge. In all cases, the inputs A and B are interchangeable. The table can be extended to gates with three or more inputs. The symbol '–' represents a 'don't care' input.

To see how the D-notation can be used, consider the circuit of Figure 10.6. To test for $A/0$, node A is first given a value D, which can be propagated via node H or via node G. To propagate the D to node H, node B must be 1. Node H then has the value $\overline{D}$.

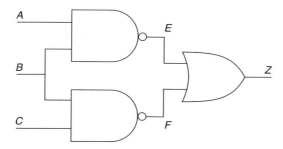

Figure 10.5   Example circuit for D-algorithm.

Table 10.1 Truth tables for the D-algorithm.

	[AND]			[OR]			[NAND]			[NOR]			[NOT]	
	A	B	Z	A	B	Z	A	B	Z	A	B	Z	A	Z
Establish fault-sensitive condition	1	1	D	0	0	$\bar{D}$	1	1	$\bar{D}$	0	0	D	1	$\bar{D}$
	0	–	$\bar{D}$	1	–	D	0	–	D	1	–	$\bar{D}$	0	D
Transmit fault	D	1	D	D	0	D	D	1	$\bar{D}$	D	0	$\bar{D}$	D	$\bar{D}$
	$\bar{D}$	1	$\bar{D}$	$\bar{D}$	0	$\bar{D}$	$\bar{D}$	1	D	$\bar{D}$	0	D	$\bar{D}$	D
Generate fixed value	1	1	1	1	–	1	1	1	0	1	–	0	1	0
	0	–	0	0	0	0	0	–	1	0	0	1	0	1
Reconvergence	D	D	D	D	D	D	D	D	$\bar{D}$	D	D	$\bar{D}$		
	$\bar{D}$	$\bar{D}$	$\bar{D}$	$\bar{D}$	$\bar{D}$	$\bar{D}$	$\bar{D}$	$\bar{D}$	D	$\bar{D}$	$\bar{D}$	D		
	D	$\bar{D}$	0	D	$\bar{D}$	1	D	$\bar{D}$	1	D	$\bar{D}$	0		

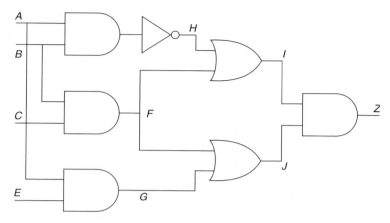

Figure 10.6   Example circuit for D-algorithm.

To propagate this $\overline{D}$ to I requires F to be 0, and to propagate the value to Z means J must be 1. If F is 0 and J is 1, G must be 1; therefore nodes A and E must both be 1. At this point we hit an inconsistency as node A has the value D. We have to return to the last decision made, which in this case was the decision to propagate the value of A through to H.

The alternative is to propagate the D at A to G. Thus, E must be 1; to propagate the value to J, F must be 0, and to propagate to Z, I must be 1. Hence H must be 1. As A is already assigned, B must be 0. This is consistent with F being 0 and C may be either 1 or 0.

The D-algorithm as presented here requires further refinement before it can be implemented as an EDA program. In particular the rules for detecting inconsistencies require more detail. Table 10.2 shows what happens when two fault-free or faulty values are propagated by different routes to the same node.

The D-algorithm is an algorithm in the true sense of the word – if a solution exists, the D-algorithm will find it. The search for a solution can, however, be very

Table 10.2   Intersection rules for the D-algorithm.

$\cap$	0	1	X	D	$\overline{D}$
0	0	ϕ	0	ψ	ψ
1	ϕ	1	1	ψ	ψ
X	0	1	X	D	$\overline{D}$
D	ψ	ψ	D	μ	λ
$\overline{D}$	ψ	ψ	$\overline{D}$	λ	μ

ϕ inconsistent logic values

ψ inconsistency between logic values and fault values

μ allowed intersection between fault values

λ inconsistent fault value

time-consuming. If necessary, every possible combination of node values will be examined. Subsequent test pattern generation algorithms have attempted to speed up the D-algorithm by improving the decision-making within the algorithm. Examples include 9-V, which uses a nine-valued algebra, and PODEM.

10.3.4 PODEM

The PODEM algorithm attempts to limit the amount of decision-making, and hence the time needed for a decision. Initially all the inputs are set to X (unknown). Arbitrary values are then assigned to the inputs and the implications of these assignments are propagated forwards. If either of the following propositions is true the assignment is rejected:

1. The node value of the fault under consideration has identical faulty and fault-free values.
2. There is no signal path from a net with a D or $\overline{D}$ value to a primary output.

We will use PODEM on the circuit of Figure 10.6 to develop a test for $H/1$. Initially, all nodes have an X value.

1. Set $A = 0$. Fails – proposition 1 (H would be 1).
2. Set $A = 1$. OK.
3. Set $B = 0$. Fails – proposition 1.
4. Set $B = 1$. OK. $H = \overline{D}$.
5. Set $C = 0$. OK. $F = 0, I = \overline{D}$.
6. Set $E = 0$. Fails – proposition 2 ($G = 0, J = 0, Z = 0$).
7. Set $E = 1$. OK. $G = 1, J = 1, Z = \overline{D}$.

Therefore a test for $H/1$ is 1101/0.

10.3.5 Fault collapsing

In the example of Figure 10.3, the test for $A/0$ (the input to a NAND gate) was the same as the test for $E/1$ (the output of that NAND gate). The same test can be used to detect $B/0$. These three faults $\{A/0, B/0, E/0\}$ are *indistinguishable*. Similarly, a test for an input of a NAND gate being stuck at 1 will also detect whether the output is stuck at 0. Two different tests are needed, however, for $A/1$ and $B/1$. Hence these faults are not indistinguishable, but an input stuck at 1 is said to *dominate* the output stuck at 0 (written $A/1 \rightarrow E/0$). The set of rules for fault indistinguishability and dominance for two-input (A, B) and single-output (Z) gates and the inverter are shown in Table 10.3.

These rules can be used to reduce a fault list. However, they do not apply to fan-out nodes, which must be omitted from any simplification procedure. If we apply these rules to the 14 faults of the circuit of Figure 10.3 we can see that we have two sets of equivalent faults: $\{A/0, B/0, E/1, F/1, Z/1\}$ and $\{C/0, D/0, F/0\}$, and the following fault dominances: $A/1 \rightarrow E/0, B/1 \rightarrow E/0, E/0 \rightarrow Z/0, F/0 \rightarrow Z/0, C/1 \rightarrow F/1$ and $D/1 \rightarrow F/1$.

Table 10.3   Fault collapsing rules.

Type of gate	Indistinguishable faults	Fault dominance
AND	{A/0, B/0, Z/0}	A/1, B/1 → Z/1
OR	{A/1, B/1, Z/1}	A/0, B/0 → Z/0
NAND	{A/0, B/0, Z/1}	A/1, B/1 → Z/0
NOR	{A/1, B/1, Z/0}	A/0, B/0 → Z/1
NOT	{A/0, Z/1}	
	{A/1, Z/0}	

As we need to test only for one fault in each equivalent set and for the dominant faults, we need to derive tests for only the following faults: $A/1$, $B/1$, $C/1$, $D/1$ and $C/0$. The fault list is cut from 14 to five faults, simplifying the fault generation task. Note that we have not lost any information by doing this – we cannot tell by observing node Z whether a fault in the circuit is one of the five listed or a fault equivalent to or dominated by one of those faults.

10.4   Fault simulation

One test pattern can be used to find more than one potential fault. For example, suppose we wish to detect whether node E is stuck at 0 in the circuit of Figure 10.7. $E/0$ dominates $G/0$ and is equivalent to $A/0$ and $B/0$. In all these cases, G will be 1 normally and 0 in the presence of one of these faults. Hence, the input pattern $A = 1$, $B = 1$, $C = 0$, $D = 0$ can be used to detect four possible faults. As there are seven nodes in the circuit, there are 14 possible stuck-at faults. This pattern covers four faults, and it can be shown that of the 16 possible input patterns, six are sufficient to detect all the possible stuck-at faults in the circuit.

It is also generally true that a fault may be covered by more than one pattern. For instance, $E/1$ can be found by attempting to force E to 0. This can be achieved by setting (a) $A = 1$, $B = 0$, (b) $A = 0$, $B = 1$, or (c) $A = 0$, $B = 0$; in all cases, $C = 0$, $D = 0$. Thus there are three possible patterns for detecting $E/1$. Note too that pattern (a) also covers

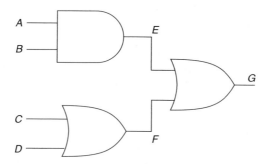

Figure 10.7   Example circuit for fault simulation.

$B/1$ and $G/1$, (b) covers $A/1$ and $G/1$, while (c) covers $G/1$. To detect all the faults in the circuit we need to use both $A = 1, B = 0, C = 0, D = 0$ and $A = 0, B = 1, C = 0, D = 0$ as these are the only patterns that detect $B/1$ and $A/1$, respectively. We are, however, applying two patterns that can detect $E/1$ and $G/1$. Having found one pattern that detects these two faults, we can *drop* the faults from further consideration. In other words, in applying the second test $A = 0, B = 1, C = 0, D = 0$, we forget about $E/1$ and $G/1$ as we already have a pattern that detects them. We could equally decide not to drop a fault when a suitable test pattern is found, in order to try to distinguish between apparently equivalent faults.

The object of fault simulation is, therefore, to assess the fault coverage of test patterns by determining whether the presence of a fault would cause the outputs of the circuit to differ from the fault-free outputs, given a particular input pattern.

The simplest approach to fault simulation is simply to modify the circuit to include each fault, one at a time, and to resimulate the entire circuit. As the single-stuck fault model assumes that only one fault can occur at a time and that each node of the circuit can be stuck at 1 and at 0, this approach, known as *serial* fault simulation, will require twice as many simulation runs as there are nodes, together with one simulation for the fault-free circuit. This technique is clearly expensive in terms of computer power and time, and three main alternatives have been suggested to make fault simulation more efficient. We will first show how these three approaches can be implemented in a simulator. In Section 10.5 we will show how a standard VHDL simulator can be used to perform fault simulation.

10.4.1   Parallel fault simulation

If we use two-state logic, one bit is sufficient to represent the state of a node. Therefore one computer word can represent the state of several nodes or the state of one node under several faulty conditions. For instance, a computer with a 32-bit word length can use one word to represent the state of a node in the fault-free circuit together with the state of the node when 31 different faults are simulated. Each bit corresponds to the circuit with one fault present. The same bit is used in each word to represent the same version of the circuit. The fault-free circuit must always be simulated, as it is important to know whether a faulty circuit can be distinguished from the fault-free circuit. If more faults are to be simulated than the number of bits in a word, the fault simulation must be completed in several passes, each of which includes the fault-free circuit.

Instead of simulating the circuit by passing Boolean values, words are used, so the state of each gate is evaluated for each fault modelled by a bit of the input signal words. Hence the name *parallel fault simulation*. Because words are passed instead of Boolean values, the event-scheduling algorithm treats any change in a word value as an event. Thus gates may be evaluated for certain versions of the circuit even if the input values for that version remain unchanged.

The circuit of Figure 10.7 has seven nodes, hence 14 possible stuck-at faults (Table 10.4). Thus 15 bits are needed for a parallel fault simulation. The word values of each node for the input pattern $A = 1, B = 1, C = 0, D = 0$ are shown below. As can

Table 10.4   Parallel fault simulation of circuit of Figure 10.7.

Bit		A	B	C	D	E	F	G
0	–	1	1	0	0	1	0	1
1	A/0	0	1	0	0	0	0	0
2	A/1	1	1	0	0	1	0	1
3	B/0	1	0	0	0	0	0	0
4	B/1	1	1	0	0	1	0	1
5	C/0	1	1	0	0	1	0	1
6	C/1	1	1	1	0	1	1	1
7	D/0	1	1	0	0	1	0	1
8	D/1	1	1	0	1	1	1	1
9	E/0	1	1	0	0	0	0	0
10	E/1	1	1	0	0	1	0	1
11	F/0	1	1	0	0	1	0	1
12	F/1	1	1	0	0	1	1	1
13	G/0	1	1	0	0	1	0	0
14	G/1	1	1	0	0	1	0	1

be seen, this pattern, as noted earlier, normally sets G to 1, but for faults $A/0$, $B/0$, $E/0$ and $G/0$, the output is 0, and therefore these faults are detected by that pattern.

There are several obvious disadvantages to parallel fault simulation. First, the number of faults that can be simulated in parallel is limited to the number of bits in a word. If more than two states are used (in other words if a state is encoded using two or more bits), the possible number of parallel faults is further reduced. As has been noted, every version of a gate is scheduled and re-evaluated whenever one of the versions of an input changes. This can be very inefficient, as a significant number of null events are likely to be processed. Moreover, if the purpose of the fault simulation is simply to detect whether any of the given test patterns will detect any of the faults, it is desirable to drop a fault from further consideration once it has proved possible to distinguish the behaviour caused by that fault from the normal, fault-free behaviour. Faults cannot be dropped in parallel fault simulation, or perhaps more accurately, the dropping of a fault is unlikely to improve the efficiency of the simulation, as the bits corresponding to that fault cannot be used for any other purpose.

10.4.2   Concurrent fault simulation

If only the differences between the fault-free simulation and the faulty simulations are maintained, constraints such as word size need not apply. On the other hand, the evaluation of gates would be made more complex because these lists of differences must be manipulated. Concurrent fault simulation maintains fault lists in the form of those gates that have different inputs and outputs in the faulty circuit from the equivalent gates in the

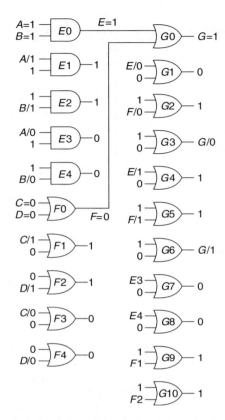

Figure 10.8 Concurrent fault simulation of the circuit of Figure 10.7.

fault-free circuit. The manipulation of fault lists thus consists of evaluating input signals, in exactly the same way as is done for the fault-free circuit, and checking to see whether the output differs from the fault-free circuit.

Figure 10.8 shows the circuit with the fault lists included for the input $A = 1, B = 1, C = 0, D = 0$. All the stuck faults for all four inputs are listed, together with the stuck faults for the internal nodes, E and F, and the output node, G. The stuck faults for E and F are listed only once. To distinguish the faulty versions of the circuit from the fault-free version, the gates are labelled according to their output nodes, together with a number. Gate 0 is always the fault-free version. A gate in the fault list is passed to a gate connected to the output only if the faulty value is different from the fault-free value. Thus $E3$, $E4$, $F1$ and $F2$ appear as inputs to gates in the fault list for G, causing faults $G7$, $G8$, $G9$ and $G10$, respectively. As with parallel fault simulation, it can be seen that for this example, $G1$, $G3$, $G7$ and $G8$, representing $E/0$, $G/0$, $A/0$ and $B/0$, respectively, have different outputs from $G0$ and are therefore detected faults.

To see why concurrent fault simulation is more efficient than parallel fault simulation, suppose that A now changes from 1 to 0. This would cause $E0$, $E2$ and $E4$ to be evaluated. $E1$ and $E3$ would not be evaluated because they both model stuck faults on A. Now, $E0$ is at 0, as are $E2$, $E3$ and $E4$; $E1$ is at 1. The OR gate, F, and its fault list would not be re-evaluated as neither C nor D changes. As faults $E3$ and $E4$ are now the same as $E0$, the corresponding faults in G ($G7$ and $G8$) are removed from the fault list and a fault corresponding to $E1$, say $G11$, is now inserted. Now gate G is evaluated, as E has changed, and faults $G2$, $G3$, $G5$, $G6$, $G9$, $G10$ and $G11$ are evaluated.

It can be seen from Figure 10.8 that, even with this small number of gates, the fault list for G has 10 elements. In practice, the fault lists can be significantly simplified with a little pre-processing of the circuit. It has already been noted that one test can cover a number of faults, and it is possible, in many cases, to deduce that some faults are indistinguishable and that tests for certain faults will always cover certain other faults. The circuit of Figure 10.8 has seven nodes and 14 stuck faults, but it can be shown that tests for only five faults ($A/0$, $C/0$, $D/0$, $A/1$ and $B/1$) are needed and that any other faults are covered by those tests. If this pre-processing is applied, faults $E4$, $F1$, $F2$, $G1$, $G2$, $G3$, $G4$, $G5$ and $G6$ can be eliminated and $G8$, $G9$ and $G10$ are in turn removed, reducing the fault list for G to one element, $G7$.

Concurrent fault simulation allows efficient selective trace and event scheduling to be used, together with the full range of state and delay models. The major disadvantage is that a significant amount of list processing must be done to propagate faults through the circuit.

10.5 Fault simulation in VHDL

Fault modelling and simulation require the perturbation of a fault-free model of a circuit. If a simulator has models of gates built into it, this perturbation can be done internally. Verilog (see Appendix B) has basic gates defined as part of the language, so a Verilog-based fault simulator can be efficiently implemented. VHDL, on the other hand, does not have gate models built in as fundamental elements. Simulators based upon VITAL (Chapter 12) can assume the existence of elementary gates. This has led to the situation where digital systems may be specified in VHDL, but faults simulated after synthesis using Verilog netlists!

Several techniques have been suggested for performing fault simulations in VHDL, by perturbing a circuit model explicitly and repeating the simulation. This can be done by including extra control wires or by including generic parameters to affect the behaviour of gates. While such perturbed netlists can be generated automatically, by writing a suitable testbench or by writing a shell program, such a technique has three major drawbacks: the netlist generated for fault simulation is not the same as that generated for fault-free simulation; the faults have to be activated by including signals or variables within the netlist; and there is no certainty that every fault will be modelled.

An alternative approach to fault simulation considered the injection of faults on wires connecting design units by changing the resolution functions of signals. This again requires the explicit listing of particular faults.

10.5.1   Fault injection

A two-input NAND gate has six possible stuck faults, but only three distinct faults need to be simulated ($A/1$, $B/1$, $Z/1$). Together with the fault-free behaviour, we therefore have four modes of behaviour. Thus a two-input NAND gate, including faulty behaviour, might be modelled as shown below, where the control signals, C0, C1 and C2, determine the mode of behaviour.

```
entity nand2 is
   port (Z : out BIT; A, B : in BIT; C0, C1, C2 : in BIT);
end entity nand2;

architecture fault_model of nand2 is
begin
  n2: process (A, B, C0, C1, C2)
  begin
    if (C0 = '1') then -- Z/1
      Z <= '1';
    elsif (C1 = '1') then -- A/1
      Z <= not B;
    elsif (C2 = '1') then -- B/1
      Z <= not A;
    else -- fault-free
      Z <= A nand B;
    end if;
  end process n2;
end architecture fault_model;
```

The control signals must be either additional input ports, as shown, or generic parameters. Thus these control signals appear in any netlist description that uses this gate model. Every control signal for every gate in a netlist must be explicitly and uniquely declared in that netlist, as below.

```
entity netlist is
   port (z : out BIT; x, y : in BIT;
         c0, c1, c2, c3, c4, c5, c6, c7, c8, c9, c10,
         c11 : in BIT);
end entity netlist;

architecture example of netlist is
   signal i0, i1, i2 : BIT;
begin
   g1 : entity WORK.nand2 port map (z, i0, i1, c0, c1, c2);
   g2 : entity WORK.nand2 port map (i0, y, i2, c3, c4, c5);
   g3 : entity WORK.nand2 port map (i1, x, i2, c6, c7, c8);
   g4 : entity WORK.nand2 port map (i2, x, y, c9, c10, c11);
end architecture example;
```

The testbench for this circuit would be complex because the control signals would have to be switched while a set of input vectors is applied.

The 1993 and 2002 VHDL standards provide an additional means of passing values to models, other than generics and ports, namely **shared variables**. Thus C0 and C1 in the NAND2 model could be shared variables whose values would be defined globally. Hence the control signals could be omitted, resulting in a netlist suitable for fault-free and fault simulation. C0 and C1 would, however, have to be unique identifiers, thus requiring every instance of a gate to be unique. This is obviously impractical.

10.5.2   Transparent fault injection

The solution shown here is to use a linked list in which each element of the list corresponds to one fault in a gate. The 2002 VHDL standard revised the way in which shared variables can be used. At the time of writing, however, there are no commercial tools available that support this standard. Here, we include a package definition that declares the necessary functions for fault simulation, compliant with the 2002 standard. Appendix C has the full package body. In addition, an equivalent package that complies with the 1993 standard is also included in Appendix C.

```vhdl
package fault_inject is
  type fault_list is protected
    impure function new_fault(name : STRING)
      return NATURAL;
    procedure first_fault;
    impure function end_fault_list return BOOLEAN;
    procedure inc_fault_list;
    impure function simulating(fault_no : NATURAL)
      return BOOLEAN;
    impure function detected return BOOLEAN;
    impure function fault_name return STRING;
    procedure set_simulate;
    procedure clr_simulate;
    procedure set_detected;
    end protected fault_list;
  shared variable fault_sim : fault_list;
end package fault_inject;

library IEEE;
use IEEE.std_logic_1164.all;
use WORK.fault_inject.all;

entity nand2 is
  port (z : out std_logic; a, b : in std_logic);
end entity nand2;
```

```vhdl
architecture inject_fault of nand2 is
begin
  nn : process (a, b) is
      variable z_sa1, a_sa1, b_sa1 : NATURAL := 0;
  begin
    if z_sa1 = 0 then
      z_sa1 := fault_sim.new_fault(
             inject_fault'INSTANCE_NAME & "z_sa1");
      a_sa1 := fault_sim.new_fault(
             inject_fault'INSTANCE_NAME & "a_sa1");
      b_sa1 := fault_sim.new_fault(
             inject_fault'INSTANCE_NAME & "b_sa1");
    end if;
    if fault_sim.simulating(z_sa1) then  -- z/1
      z <= '1';
    elsif fault_sim.simulating(a_sa1) then  -- a/1
      z <= not b;
    elsif fault_sim.simulating(b_sa1) then  -- b/1
      z <= not a;
    else  -- fault-free
      z <= a nand b;
    end if;
  end process nn;
end architecture inject_fault;
```

Package fault_inject contains the definition of a protected type. In this protected type, a number of functions are declared for manipulating the data structure.

The package body contains the body of the protected type. A record of type
fault_model is created for *each* fault, containing the name of the fault, a
Boolean flag indicating whether that fault is being simulated, a second Boolean flag
to indicate whether any test has detected that potential fault, and a pointer to the
next fault. Each fault has an index. The first fault in the list is indexed by the
shared variable.

The code for a two-input NAND gate is shown above. Within each gate model, its
fault records are created at the beginning of a simulation run. From Chapter 8, it should
be recalled that at initialization, every process is executed once until it suspends.
A local variable indexes each fault (z_sa1, a_sa1, b_sa1), so that it can be identified
during a simulation. If any of these local variables is 0, the model has not yet been executed, so the data structure is created. Note that the order of evaluation of gate models
is indeterminate, as the order of execution of processes is indeterminate, but this does
not matter here.

The method fault_sim.new_fault creates the data structure for each fault.
The name of the fault is generated using the INSTANCE_NAME attribute, appending a
string to represent the fault. The INSTANCE_NAME attribute gives the entire path
name through the hierarchy of an architecture or entity, allowing each fault to be
uniquely identified. It should be noted that the entity declaration of this gate model

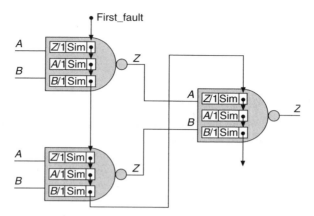

Figure 10.9 Fault list constructed in VHDL model.

contains no reference to the fault models contained in the gate, thus such a gate model can be used in any netlist with no modification to that netlist, as shown in the following model of a full adder. Figure 10.9 shows how the fault list might be constructed around a structural description.

The gate models with fault injection can be used in a standard netlist. The full adder, below, uses inverters and two-, three- and four-input NAND gates. The models of the inverter and the three- and four-input NAND gates are written in exactly the same way as that of the two-input NAND gate above. Note that there is no reference to the fault injection package, nor to the particular gate models in the full adder model.

```
library IEEE;
use IEEE.std_logic_1164.all;

entity FullAdder is
  port (x, y, Cin : in std_logic;
        Sum, Cout: out std_logic);
end entity FullAdder;

architecture FA of FullAdder is
  signal c0, c1, c2, i0, i1, i2, s0, s1, s2, s3 : std_logic;
begin
  g1  : entity WORK.nand3 port map (Cout, c0, c1, c2);
  g2  : entity WORK.nand2 port map (c0, x, y);
  g3  : entity WORK.nand2 port map (c1, x, Cin);
  g4  : entity WORK.nand2 port map (c2, y, Cin);
  g5  : entity WORK.nand4 port map (Sum, s0, s1, s2, s3);
  g6  : entity WORK.inv port map (i0, x);
  g7  : entity WORK.inv port map (i1, y);
  g8  : entity WORK.inv port map (i2, Cin);
  g9  : entity WORK.nand3 port map (s0, x, i1, i2);
  g10 : entity WORK.nand3 port map (s1, i0, y, i2);
```

```
g11 : entity WORK.nand3 port map (s2, i0, i1, Cin);
g12 : entity WORK.nand3 port map (s3, x, y, Cin);
end architecture FA;
```

10.5.3   VHDL fault simulation

The other part of the procedure is the fault simulation. In this example, we will set up a fault simulation for a four-bit ripple adder, as shown in Section 4.5.2. The code below shows the structure of part of a testbench. The fault-free simulation is performed first and the fault-free responses obtained. The fault simulations are then performed in sequence. A pointer is moved along the list of faults. As it points to each fault record, a flag in each record is set true to activate that fault and the test stimuli are applied. The responses can then be compared with the fault-free responses. Although the simulations are performed sequentially and hence the absolute time increases, it is easy to calculate the relative time at which each vector is applied.

```
sim : process is
begin
  -- FAULT-FREE SIMULATION
  -- apply test vectors
  -- apply Xs
  -- SEQUENTIAL FAULT SIMULATION
  fault_sim.first_fault;
  while not fault_sim.end_fault_list loop
    fault_sim.set_simulate;
    -- apply test vectors
    -- compare with fault free case and print differences
    fault_sim.clr_simulate;
    -- apply Xs
    -- move to next fault:
    fault_sim.inc_fault_list;
  end loop;
end process sim;
```

The testbench, below, takes the test vectors from a file (vectors.txt) and generates a file of correct responses, and a summary file, showing which vectors detect which faults. The vectors file is of the form:

```
0000 0000 0
0000 0001 0
0101 0101 0
1111 1111 0
1111 1111 1
```

Note that after the final vector is applied, all the inputs are set to the unknown value ('X') in order to force a re-evaluation of all gates when the next fault is activated. If this

were not done, certain faults might be missed, because the state of nodes could be
unchanged between the last vector for one fault and the first vector for the next fault.
Hence signals in the sensitivity lists would be unaffected, even though the flag in the
fault list had been moved. Therefore the gates would otherwise not be re-evaluated.

```vhdl
library IEEE;
use IEEE.std_logic_1164.all, STD.textio.all,
    WORK.fault_inject.all;

entity tb is
end entity tb;

architecture fileio of tb is
  file vectors : text;
  file results : text;
  file faults : text;
  constant N : NATURAL := 4;
  signal X, Y, Z: std_logic_vector(N-1 downto 0)
                  := (others => '0');
  signal ci, co: std_logic := '0';
begin
  a1: entity WORK.NBitAdder(StructIterative)
        port map (X, Y, ci, Z, co);
  p1: process is
    variable ILine, OLine, fname : Line;
    variable ch : CHARACTER;
    variable fc, fd : NATURAL := 0;
    variable rel_time, abs_time : TIME;
    variable X_in, Y_in, Z_in: BIT_VECTOR(N-1 downto 0);
    variable ci_in, co_in: BIT;
  begin
    file_open(vectors, "vectors.txt", READ_MODE);
    file_open(results, "results.txt", WRITE_MODE);
    while not endfile(vectors) loop
        readline(vectors, ILine);
        read(ILine, X_in);
        read(ILine, ch);
        read(ILine, Y_in);
        read(ILine, ch);
        read(ILine, ci_in);
        X <= to_stdlogicvector(X_in);
        Y <= to_stdlogicvector(Y_in);
        ci <= to_stdulogic(ci_in);
        wait for 100 NS;
        write(OLine, X_in, right, 5);
        write(OLine, Y_in, right, 5);
```

```
        write(OLine, ci_in, right, 2);
        write(OLine, to_bitvector(Z), right, 5);
        write(OLine, to_bit(co), right, 2);
        writeline(results, OLine);
      end loop;
      file_close(vectors);
      file_close(results);
--    Force circuit into unknown state
      X <= (others => 'X');
      Y <= (others => 'X');
      ci <= 'X';
      wait for 100 NS;
      abs_time := NOW;
      fault_sim.first_fault;
      file_open(faults, "faults.txt", WRITE_MODE);
      while not fault_sim.end_fault_list loop
        fc := fc + 1;
        fault_sim.set_simulate;
        file_open(results, "results.txt", READ_MODE);
        while not endfile(results) loop
          readline(results, ILine);
          read(ILine, X_in);
          read(ILine, ch);
          read(ILine, Y_in);
          read(ILine, ch);
          read(ILine, ci_in);
          read(Iline, ch);
          read(Iline, Z_in);
          read(Iline, ch);
          read(Iline, co_in);
          X <= to_stdlogicvector(X_in);
          Y <= to_stdlogicvector(Y_in);
          ci <= to_stdulogic(ci_in);
          wait for 100 NS;
          if (to_stdlogicvector(Z_in) /= Z or
            to_stdulogic(co_in) /= co) then -- fault detected
            fault_sim.set_detected;
            fname := new STRING'(fault_sim.fault_name);
            writeline(faults, fname);
            write(OLine, STRING'("Fault #"), left, 0);
            write(OLine, fc, left, 5);
            write(OLine, STRING'("Detected by input:"),left,0);
            write(OLine, X_in, right, 5);
            write(OLine, Y_in, right, 5);
            write(OLine, ci_in, right, 2);
```

```
                write(OLine, STRING'(" output: "), left, 0);
                write(OLine, to_bitvector(Z), right, 5);
                write(OLine, to_bit(co), right, 2);
                write(OLine, STRING(" expected: "), left, 0);
                write(OLine, Z_in, right, 5);
                write(OLine, co_in, right, 2);
                write(OLine, STRING'(" at "), left, 0);
                rel_time := NOW — fc*abs_time;
                write(OLine, rel_time, right, 9);
                write(OLine, NOW, right, 9);
                writeline(faults, OLine);
            end if;
        end loop;
        fault_sim.clr_simulate;
-- Force circuit into unknown state
        X <= (others => 'X');
        Y <= (others => 'X');
        ci <= 'X';
        wait for 100 NS;
        fault_sim.inc_fault_list;
        file_close(results);
      end loop;
-- summarize results
    fault_sim.first_fault;
    while not fault_sim.end_fault_list loop
      if fault_sim.detected then
        fd := fd + 1;
      end if;
      fault_sim.inc_fault_list;
    end loop;
    write(OLine, STRING'(" Fault Cover:"), left, 0);
    writeline(faults, OLine);
    write(OLine, fc, right, 8);
    write(OLine, STRING'(" faults, "), left, 0);
    write(OLine, fd, right, 8);
    write(OLine, STRING'(" detected "), left, 0);
    writeline(faults, OLine);
    wait; -- halt
  end process p1;
end architecture fileio;
```

This generates an output file of the form:

```
:tb(fileio):al@nbitadder(structiterative):g1(0):lt:f0@
fulladder(fa):g1@nand_n(inject_fault):z_sa1
Fault #4 Detected by input: 0000 0000 0 output: 0010 0
expected: 0000 0 at 100 ns 2100 ns
:tb(fileio):al@nbitadder(structiterative):g1(0):lt:f0@
fulladder(fa):g1@nand_n(inject_fault):z_sa1
Fault #4 Detected by input: 0000 0001 0 output: 0011 0
expected: 0001 0 at 200 ns 2200 ns
:tb(fileio):al@nbitadder(structiterative):g1(0):lt:f0@
fulladder(fa):g2@nand_n(inject_fault):a(1)_sa1
Fault #6 Detected by input: 0000 0001 0 output: 0011 0
expected: 0001 0 at 200 ns 3200 ns
:tb(fileio):al@nbitadder(structiterative):g1(0):lt:f0@
fulladder(fa):g2@nand_n(inject_fault):z_sa1
Fault #7 Detected by input: 0101 0101 0 output: 1000 0
expected: 1010 0 at 300 ns 3800 ns
Fault Cover:
160 faults, 60 detected
```

In the testbench, notice that all the file-handling functions are used. The vectors are read as `bit_vectors` and converted to `std_logic_vectors`. A package, `std_logic_textio`, can be found on the Internet for reading and writing `std_logic` types. It does not offer any particular advantage here. `Read` must have a variable as its output parameter, so a signal assignment must be performed.

When results are written out to the file `faults.txt`, each line is first written to the line buffer `OLine`. The `writeline` procedure then writes out the line buffer and deallocates the buffer. The line

```
fname := new STRING'(fault_sim.fault_name);
```

copies the name of each fault to a new string. This copy is then written out and deallocated. If this is not done, i.e. if we wrote

```
writeline(faults, fault_sim.fault_name);
```

the name in the data structure would be deallocated.

Finally, the forms of the **loop** statement should be noted. The number of vectors in the input file and the number of faults can be varied without modifying the testbench code.

Summary

The principles of digital testing have been introduced. Defects are characterized as logical faults. Test pattern generation algorithms have been described. Parallel and concurrent fault simulation algorithms have also been discussed.

A VHDL implementation of a sequential fault simulator has been described. This includes a number of advanced VHDL features: pointers, string handling, and file input and output.

Further reading

Abramovici, Breuer and Friedman is a very good introduction to fault modelling, test generation and fault simulation. Also recommended are the books by Wilkins and Miczo. New fault models and algorithms are still being developed, with particular emphasis on delay effects and on sequential systems. *IEEE Design and Test of Computers* provides a quarterly update on developments.

Exercises

10.1 Explain the difference between structural and functional testing.

10.2 What assumptions are made by the single-stuck fault model?

10.3 Write down the stuck-at-fault list for the circuit shown in Figure 10.10. Derive tests for $A/1$ and $A/0$ and determine which other faults these tests cover. Show that it is not possible to derive a test for $G/0$.

10.4 Suggest a test pattern to determine whether nodes H and I in Figure 10.10 are bridged together. You should assume that a bridging fault may be modelled as a wired-OR; i.e. that if either wire is at logic 1, the other wire is also pulled to a logic 1.

10.5 A positive edge-triggered D-type flip-flop is provided with an active-low asynchronous clear input, and has only its Q output available. By considering the functional behaviour of the flip-flop, develop a test sequence for this device for all single-stuck faults on inputs and outputs.

10.6 Describe the four types of crosspoint fault that can occur in a PLA consisting of an AND plane and an OR plane.

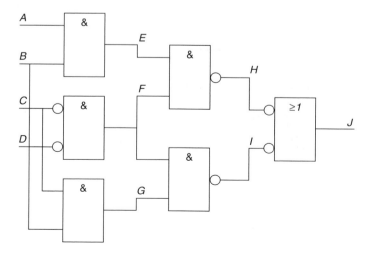

Figure 10.10 Circuit for Exercises 10.3 and 10.4.

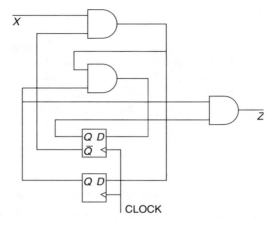

Figure 10.11   Circuit for Exercise 10.8.

10.7   The AND and OR planes of a PLA can be thought of as two NAND planes. What is the minimal set of test patterns required to test an n-input NAND gate?

10.8   Write down a stuck-fault list for the circuit in Figure 10.11. How, in principle, would a test sequence for this circuit be constructed?

10.9   The circuit shown in Figure 10.12 is an implementation of a state machine with one input and one output. Derive the next state and output equations and hence show that a parasitic state machine exists, in addition to the intended state machine. Assuming that the initial state of the flip-flops is $P = Q = 0$, suggest

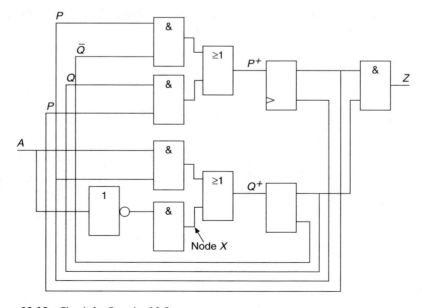

Figure 10.12   Circuit for Exercise 10.9.

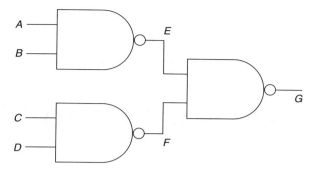

Figure 10.13   Circuit for Exercise 10.11.

a sequence of input values at *A* that will cause the output, *Z*, to have the values 0, 1, 1, 0 on successive clock cycles. Hence, show that this sequence of input values can be used to test whether node *X* is stuck at 0.

10.10 Explain the difference between *parallel* and *concurrent* fault simulation.

10.11 In the circuit of Figure 10.13, *A* = 1, *B* = 1, *C* = 1 and *D* = 0. Derive the fault lists as they would be included in a concurrent fault simulator, assuming that each of the nodes can be stuck at 1 or stuck at 0. Show that the fault lists may be significantly simplified if redundant and dominated faults are removed in a pre-processing step.

Chapter 11

Design for testability

As noted in the previous chapter, testability for a circuit such as that shown in Figure 11.1 can be expressed in terms of:

- Controllability – the ability to control the logic value of an internal node from a primary input.
- Observability – the ability to observe the logic value of an internal node at a primary output.

The previous chapter discussed methods for finding test patterns for combinational circuits. The testing of sequential circuits is much more difficult because the current state of the circuit as well as its inputs and outputs must be taken into account. Although in many cases it is possible, at least in theory, to derive tests for large, complex, sequential circuits, in practice it is often easier to modify the design to increase its testability. In other words, extra inputs and outputs are included to increase the controllability and observability of internal nodes.

Testability can be enhanced by *ad hoc* design guidelines or by a structured design methodology. In this chapter we shall discuss general *ad hoc* principles for increasing testability, then look at a structured design technique – the scan path. In the third section, we will see how some of the test equipment itself can be included on an integrated circuit to provide self-test capabilities. Finally, the scan path principle can be used for internal testing, but it can also be used to test the interconnect between integrated circuits – boundary scan.

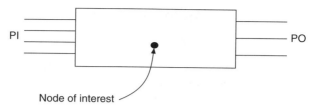

Figure 11.1 Testability of a node.

(11.1) *Ad hoc* testability improvements

If one of the objectives of a design is to enhance the testability of that design, there are a number of styles of design that should be avoided, including:

- Redundant logic. As seen in the previous chapter, redundant combinational logic will result in potentially undetectable faults. This means that the design is not fully testable and also that time may be spent attempting to generate tests for these undetectable faults.

- Asynchronous sequential systems (and in particular unstructured asynchronous systems) are difficult to synchronize with a tester. The operation of a synchronous system can be halted with the clock. An asynchronous system is, generally, uncontrollable. If asynchronous design is absolutely necessary, confine it to independent blocks.

- Monostables are sometimes used for generating delays. They are extremely difficult to control, and again should be avoided.

On the other hand, there are a number of modifications that could be made to circuits to enhance testability. The single most important of these is the inclusion of some form of initialization. A test sequence for a sequential circuit must start from a known state. Therefore initialization must be provided for all sequential elements, as shown in Figure 11.2. Any defined state will do – not necessarily all zeros. Multiple initial states can be useful.

The cost of enhancing testability includes that of extra I/O pins (including interfaces, etc.), extra components (MUXs), extra wiring, and the degradation of performance because of extra gates in signal paths; in general, there are more things to go wrong. Against this must be set the benefit that the circuit will be easier to test and hence the manufacturer and consumer can be much more confident that working devices are being sold.

(11.2) Structured design for test

The techniques described in the previous section are all enhancements that can be made to a circuit after it has been designed. A structured design for test method should consider the testability problem from the beginning. Let us restate the problem to see how it can be tackled in a structured manner.

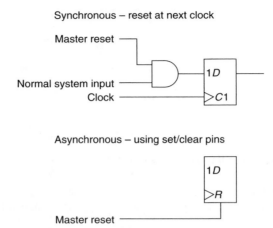

Figure 11.2   Resets add testability.

Testing combinational circuits is relatively easy provided there is no redundancy in the circuit. The number of test vectors is (much) less than $2^{(no.\ of\ inputs)}$. Testing sequential circuits is difficult because such circuits have states. A test may require a long sequence of inputs to reach a particular state. Some faults may be untestable, because certain states cannot be reached. Synchronous sequential systems, however, can be thought of as combinational logic (next state and output logic) and sequential logic (registers). Therefore the key to structured design for test is to separate these two elements.

A synchronous sequential system does not, however, provide direct control of all inputs to the combinational logic, does not allow direct observation of all outputs from the combinational logic, and does not allow direct control or observation of the state variables.

The scan-in, scan-out (SISO) principle overcomes these problems by making the state variables directly accessible by connecting all the state registers as a shift register, for test purposes, as shown in Figure 11.3. This shift register has a mode control input, M. In normal, operational mode, M is set to 0. In scan mode, M is set to 1 and the flip-flops form a shift register, the input to the shift register being the scan data in (SDI) pin and the output being the scan data out (SDO) pin.

If the combinational logic has no redundancies, a set of test patterns can be generated for it, as if it were isolated from the state registers. The test patterns and the expected responses then have to be sorted because this test data is applied through the primary inputs and through the state registers using the scan path. Similarly, the outputs of the combinational logic are observed through the primary outputs and using the scan path.

The scan path is used to test a sequential circuit using the following procedure.

1. Set $M = 1$ and test the flip-flops as a shift register. If a sequence of 1s and 0s is fed into SDI, we would expect the same sequence to emerge from SDO delayed by the number of clock cycles equal to the length of the shift register (n). A useful test sequence would be 00110 . . . which tests all transitions and whether the flip-flops are stable.

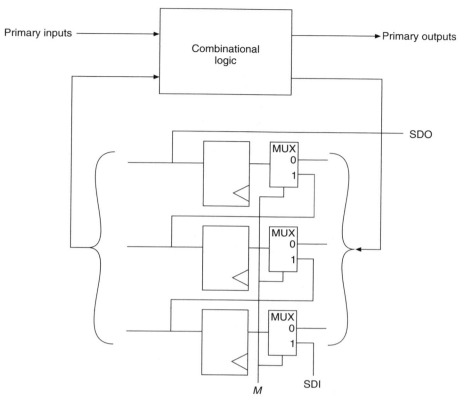

Figure 11.3 SISO principle.

2. Test the combinational logic.

    (a)   Set $M = 1$ to set the state of the flip-flops after n clock cycles by shifting a pattern in through SDI.

    (b)   Set $M = 0$. Set up the primary inputs. Collect the values of the primary outputs. Apply one clock cycle to load the state outputs into the flip-flops.

    (c)   Set $M = 1$ to shift the flip-flop contents to SDO after $n - 1$ clock cycles.

Note that step 2(a) for the next test can be done simultaneously with step 2(c) for the present test. In other words, while the contents of the shift register are being shifted out, new data can be shifted in behind it.

The benefit of using a scan path is that it provides an easy means of making a sequential circuit testable. If there is no redundancy in the combinational logic, the circuit is fully testable. The problem of test pattern generation is reduced to generating tests only for the combinational logic. This can mean that the time to test one device can be greater than would be the case if specific sequential tests had been generated.

The costs of SISO include extra hardware: at least one extra pin for M; SDI and SDO can be shared with other system functions by using multiplexers. An extra multiplexer is needed for each flip-flop and extra wiring is needed for the scan path. Hence this can lead to performance degradation as the delay through the next state logic is increased. To minimize the wiring, it makes sense to decide the order of registers in the scan path *after* placement of devices on an ASIC or FPGA has been completed. The order of registers is unimportant provided it is known to the tester.

SISO has now become relatively well accepted as a design methodology. Most VLSI circuits include some form of scan path, although this is not usually documented.

A number of variations to SISO have been proposed including multiple scan paths – put flip-flops in more than one scan path to shorten the length of each path and to shorten the test time – and partial scan paths, whereby some flip-flops are excluded from the scan path.

(11.3) Built-in self-test

As with all testing matters, the motivation for built-in self-test (BIST or BIT for built-in test) is economic. The inclusion of structures on an integrated circuit or board that not only enhance the testability but also perform some of the testing simplifies the test equipment and hence reduces the cost of that equipment. BIST can also simplify test pattern generation because the test vectors are generated internally, and it allows field testing to be performed for perhaps years after manufacture. Overall, therefore, BIST should increase user confidence.

The principle of BIST is shown in Figure 11.4. The test vector generation and checking are built on the same integrated circuit as the circuit under test. Thus there are two obvious problems: how to generate the test vectors and how to check the responses. It would, in principle, be possible to store pre-generated vectors in ROM. There could, however, be a very large number of vectors. Similarly, it would be possible to have a look-up table of responses.

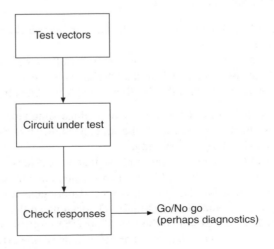

Figure 11.4   BIST principle.

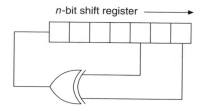

Figure 11.5 LFSR.

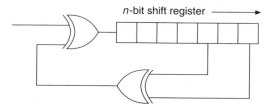

Figure 11.6 SISR.

If an exhaustive test were conducted, all possible test vectors could be generated using a binary counter. This could require a substantial amount of extra combinational logic. A simpler solution is to use a linear feedback shift register (LFSR), introduced in Chapter 6. An LFSR is a pseudo-random number generator that generates all possible states (except the all 0s state) but requires less hardware than a binary counter, as shown in Figure 11.5.

A similar structure can be used instead of a look-up table to collect the responses. The single-input signature register is shown in Figure 11.6. This is an LFSR with a single data input. The register holds the residue from a modulo-2 division. In other words, it compresses the stream of input data to produce a signature that may be compared, after a certain number of cycles, with a known good signature.

Another variant is the multiple input signature register (MISR), shown in Figure 11.7. Again, this is a modified LFSR but with more than one data input. Thus, a number of output vectors can be gathered and compressed. After a number of clock cycles the

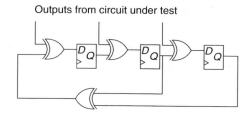

Figure 11.7 MISR.

signature in the register should be unique. If the circuit contains a fault, the register should contain an incorrect signature, which can easily be checked.

This approach will obviously fail if the MISR is sensitive to errors. The probability that a faulty circuit generates a correct signature tends to 2^{-n} for an n-stage register and long test sequences.

11.3.1 Example

For example, consider a circuit consisting of three parts: a three-stage LFSR, a three-stage MISR and the circuit under test, with the following functions:

$$X = A \oplus B \oplus C$$
$$Y = A.B + A.C + B.C$$
$$Z = \overline{A}.B + \overline{A}.C + B.C$$

The structure of the circuit is shown in Figure 11.8.

In order to see what the correct signature should be, we can perform a simulation. A VHDL model of an LFSR was presented in Chapter 6. This model can easily be adapted to implement an MISR (see the exercises at the end of this chapter). The circuit under test can be described in VHDL by the following model.

```vhdl
library IEEE;
use IEEE.std_logic_1164.all;

entity CUT is
    port(a_in, b_in, c_in : in std_logic;
         x, y, z : out std_logic);
end entity CUT;

architecture Fault_Model of CUT is
    signal a, b, c: std_logic;
begin
    a <= a_in;
    b <= b_in;
    c <= c_in;
    x <= a xor b xor c;
    y <= (a and b) or (a and c) or (b and c);
    z <= (not a and b) or (not a and c) or (b and c);
end architecture Fault_Model;
```

The input signals a_in, b_in and c_in are not used directly because we will insert fault models into those signals later. The testbench for this circuit can therefore consist of the following code.

```vhdl
library IEEE;
use IEEE.std_logic_1164.all;

entity bistex is end entity bistex;
```

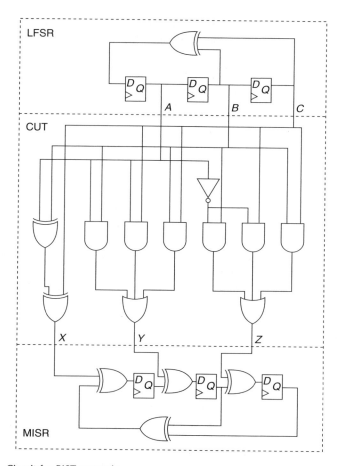

Figure 11.8 Circuit for BIST example.

```
architecture tb of bistex is
  constant n : NATURAL := 3;
  signal clock, reset : std_logic := '0';
  signal signature, q, z : std_logic_vector(n-1 downto 0);
begin
  l0: entity WORK.lfsr generic map (n)
                       port map (reset, q, clock);
  m0: entity WORK.misr generic map (n)
                       port map (reset, z, signature,
                                 clock);
  c0: entity WORK.CUT port map (q(2), q(1), q(0), z(2),
     z(1), z(0));
  reset <= '0', '1' after 5 NS, '0' after 10 NS;
  clock <= not clock after 20 NS;
end architecture tb;
```

Both the LFSR and MISR are initialized to the 111 state. When the VHDL model is simulated, we get the following sequence of states:

LFSR output	CUT output	MISR
abc	xyz	
111	111	111
011	011	100
001	101	001
100	100	001
010	101	000
101	010	101
110	010	100
111	111	000

The last output of the MISR, 000, is the signature of the fault-free circuit. The intermediate values of the MISR are irrelevant.

We can emulate a stuck fault at the input by changing one of the assignment statements in the CUT. To model a stuck-at-0, the line

```
a <= a_in;
```

is changed to

```
a <= '0';
```

(We could, of course, implement a full fault simulation model, as described in the previous chapter.) If this perturbed circuit is simulated, the sequence of states is now:

LFSR output	CUT output	MISR
abc	xyz	
111	011	111
011	011	000
001	101	011
100	000	100
010	101	010
101	101	000
110	101	101
111	011	011

The signature of the circuit when a is stuck at 0 is therefore 011. We do not care about the sequence of intermediate states. Hence a comparison of the value in the MISR with 000 when the LFSR is at 111 would provide a pass/fail test of the circuit. In principle, we could simulate every fault in the circuit and note its signature. This information could be used for fault diagnosis. In practice, of course, we would be assuming that every defect manifests itself as a single stuck fault, so this diagnostic information would have to be used with some caution. Moreover, both the LFSR and MISR could themselves contain faults, which in turn would generate incorrect signatures.

If we run the simulation again for a stuck-at-1, the signature 000 is generated. This is an example of *aliasing* – a fault generates the same signature as the fault-free circuit. The probability of aliasing can be shown to tend to 2^{-n} if a maximal length sequence is used. As there are only three stages to the MISR, the probability of aliasing is 2^{-3} or 1/8. With larger MISRs the probability of aliasing decreases.

In this example, we have made the LFSR and the MISR the same size and used the complete sequence of inputs once. None of these restrictions is essential. We can use LFSRs of different lengths and we do not need to use all the outputs from the LFSR nor all the inputs to the MISR. We can use a shorter sequence than the complete cycle of the LFSR or we can run through the sequence more than once. In all cases, however, the sequence has to be defined when the circuit is built.

11.3.2   Built-in logic block observation (BILBO)

The LFSR and MISR, described above, are specialist logic blocks. To include BIST in a circuit using such blocks would require additional registers to those required for normal operation. A scan path reuses the existing registers in a design for testing; in much the same way, built-in logic block observation (BILBO) registers are used both for normal operation and for BIST. A typical BILBO architecture is shown in Figure 11.9. Three control signals are required, which control the circuit as follows.

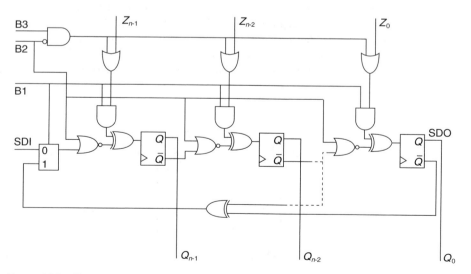

Figure 11.9   BILBO.

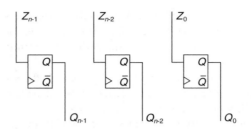

Figure 11.10　BILBO in normal mode.

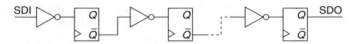

Figure 11.11　BILBO in scan mode.

B1	B2	B3	Mode
1	1	–	Normal
0	1	–	Reset
1	0	0	Signature analysis MISR
1	0	1	Test pattern generation LFSR
0	0	–	Scan

To understand the functionality of the circuit, it helps to redraw the functionality of the BILBO when the control signals are set to their different states. Figures 11.10, 11.11 and 11.12 show the normal mode, scan mode and LFSR/MISR modes respectively. Note that in the scan, LFSR and MISR modes, the Q output of the flip-flops is used, but is inverted before being fed into the next stage. The reset mode synchronously initializes the flip-flops to 0. It was noted in Chapter 6 that an LFSR stays in the all-0s state if it ever enters that state. In LFSR/MISR modes, the BILBO inverts the feedback

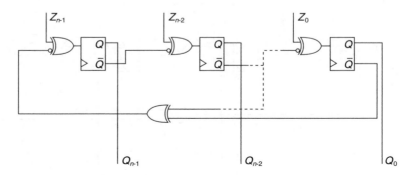

Figure 11.12　BILBO in LFSR/MISR mode.

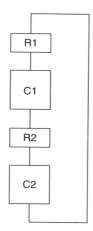

Figure 11.13   Circuit partitioning for self-test.

signal, thus making the all-0s state valid, but there still remain $2^n - 1$ states in the cycle – one state is excluded from the normal sequence.

Unlike the flip-flops in a scan path, the flip-flops in a BILBO-oriented system must be grouped into discrete registers. (The scan mode also allows us to link all the BILBOs in a scan path – see below.) These registers would ideally replace the normal system registers. An example of a system using BILBOs for self-test is shown in Figure 11.13. R1 and R2 are BILBOs, and C1 and C2 are blocks of combinational logic. To test C1, R1 is configured as an LFSR, and R2 is configured as an MISR. Similarly, to test C2, R2 is configured as an LFSR, and R1 as an MISR.

A different arrangement is shown in Figure 11.14. R1, R2 and R3 are BILBOs; C1, C2 and C3 are combinational logic. To test C1, R2 is an LFSR and R1 is an MISR. To test C2, R1 is an LFSR and R2 is an MISR; and so on.

We can therefore use BILBOs to test different structures of combinational logic, but we also need to have some confidence in the correct operation of the BILBOs themselves. Thus, how do we test the BILBOs? The first act in any test must be initialization. This can be done using the synchronous reset. Then the scan path can be used to

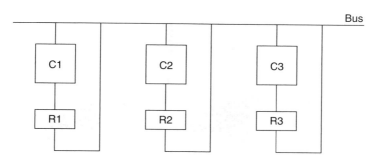

Figure 11.14   Alternative circuit partitioning for self-test.

test the flip-flops. This implies that some form of controller is needed to generate the BILBO control signals. It is not possible to test that controller (because a further controller would be needed, which in turn would need to be tested, *ad infinitum*). Therefore some form of reliable controller is needed to oversee the self-test regime. It makes sense therefore to adopt a 'start small' strategy, in which part of the system is verified before being used to test a further part of the system. If the system includes some form of microprocessor, software-based tests can be performed once the microprocessor has been checked.

Before adopting BIST in a design, the cost and effectiveness of the strategy must be considered. There is, of course, the cost of additional hardware – just over four gates per flip-flop for a BILBO-based design, together with the cost of a test controller and the additional assorted wiring. This means that there will be an increased manufacturing cost. The extra hardware means that the reliability of the system will be decreased – there is more to go wrong. There is also likely to be some performance degradation as the hardware between flip-flops is increased. The incorporation of BIST means that the complexity of the design and hence the time taken to do the design is increased. On the other hand, using BIST means that the costs of test pattern generation disappear and that the equipment needed to test integrated circuits can be simplified. Moreover the tests can be performed every time the circuit is switched on, not merely once at the time of manufacture.

(11.4) Boundary scan (IEEE 1149.1)

The techniques described so far in this chapter have been oriented towards integrated circuits, in which controllability and observability may be limited. Circuits built from discrete gates on printed circuit boards (PCBs) are generally considered easier to test because it is possible to gain access to all the nodes of the circuit using a probe, as shown in Figure 11.15, or a number of probes arranged as a 'bed-of-nails'. This assumption has become invalid in recent years for the following reasons:

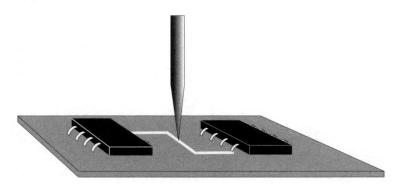

Figure 11.15   Probe testing.

- It is not possible to test mounted ICs (the pins may be connected together).
- PCBs now often have more than 20 layers of metal, so deep layers cannot be reached.
- The density of components on a PCB is increasing. Multi-chip modules (MCMs) take the chip/board concept further and have unpackaged integrated circuits mounted directly on a silicon substrate.

Boundary scan is a technique for testing the interconnect on PCBs and for testing ICs mounted on PCBs. As before, both the ICs and the empty PCB can be tested, but boundary scan replaces the step of testing the loaded PCB with a 'bed-of-nails' tester. The bed-of-nails approach has also been criticized because of 'backdriving' – in order to test a single gate its inputs would be forced to particular logic values, which also forces those logic values onto the outputs of other gates. This is not how gates are designed to work and may cause them damage.

The principle of boundary scan is to allow the *outputs* of each IC to be controlled and *inputs* to be observed. For example, consider the faults shown in Figure 11.16. These faults are external to the integrated circuits and have arisen as a result of assembling (fault-free) ICs onto a PCB. Instead of using mechanical probes to access the board, the faults are sensitized electrically. The outputs of the left-hand ICs in Figure 11.16 are used to establish test patterns and the inputs of the right-hand IC are used to observe the responses. Therefore we need to control and observe the output and input pins, respectively, of the integrated circuits. This can be done by connecting those pins on the *boundary* of the integrated circuits into a *scan* path, using special logic blocks at each input and output.

Figure 11.17 shows how the input and output pins of all the integrated circuits on a board are connected together in a scan path. Each IC has dedicated pins to allow the scan path to pass through it. These pins are labelled as TDI (Test Data In) and TDO (Test Data Out). In addition, control pins will be needed. The various ICs on a board may come from different manufacturers. For boundary scan to work, the ICs need to use the same protocols. Therefore an IEEE standard, 1149.1, has been defined. This

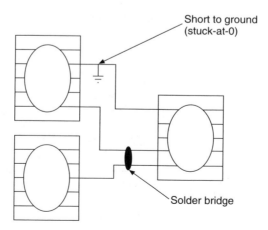

Figure 11.16   Circuit board faults.

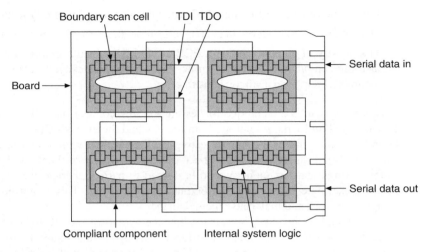

Figure 11.17   Board with boundary scan.

standard arose from the work of the Joint Test Action Group (JTAG). The term JTAG is therefore often used in reference to the boundary scan architecture.

Every boundary scan-compliant component has a common test architecture, shown in Figure 11.18. The elements of this architecture are as follows.

1. Test access port (TAP)

The TAP consists of four or five additional pins for testing. The pins are:

● TDI and TDO (Test Data In and Test Data Out). Both data and instructions are sent to ICs through the scan path. There is no way to distinguish data from instructions, or indeed to determine which particular IC a sequence of bits is intended to reach. Therefore the following pin is used to control where the data flows.

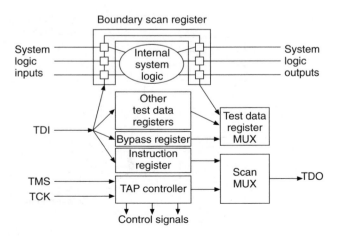

Figure 11.18   Boundary scan architecture.

● TMS (Test Mode Select). Together with the TCK pin, the TMS pin is used to control a state machine that determines the destination of each bit arriving through TDI.

● TCK (Test Clock).

● TRST (Test Reset) which is an optional asynchronous reset (not shown in Figure 11.18).

2. TAP controller

This is a 16-state machine that controls the test. The inputs to the state machine are TCK and TMS. The outputs are control signals for other registers. The state chart of the TAP controller is shown in Figure 11.19. Notice that a sequence of five 1s on TMS in successive clock cycles will put the state machine into the Test-Logic-Reset state from any other state. The control signals derived from the TAP controller are used to enable other registers in a device. Thus a sequence of bits arriving at TDI can be sent to the instruction register or to a specific data register, as appropriate.

3. Test data registers

A boundary scan-compliant component must have all its inputs and outputs connected into a scan path. Special cells, described below, are used to implement the scan register. In addition, there must be a bypass register of one bit. This allows the scan path to be shortened by avoiding the boundary scan register of a component.

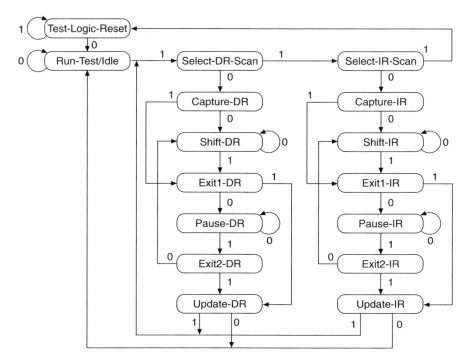

Figure 11.19 TAP controller state diagram.

Other registers may also be included; for example, an IC might include an identification register, the contents of which could be scanned out to ensure that the correct device had been included on a PCB. Similarly, the internal scan path of a device could be made accessible through the boundary scan interface. Some programmable logic manufacturers allow the boundary scan interface to be used for programming devices. Thus the configuration register is another possible data register.

4. Instruction register

This register has at least two bits, depending on the number of tests implemented. It defines the use of test data registers. Further control signals are derived from the instruction register.

The core logic is the normal combinational and sequential logic of the device. This core logic may (should) contain a scan path and may also contain BIST structures.

A typical boundary scan cell is shown in Figure 11.20. This cell can be used for an input or an output pin. For an input pin, IN is connected to the pin, and OUT is connected to the device core; for an output pin, IN comes from the core, and OUT goes to the pin. Other designs of boundary scan cell are possible.

The boundary scan cell has four modes of operation.

1. Normal mode. Normal system data flows from IN to OUT.

2. Scan mode. ShiftDR selects the SCAN_IN input; ClockDR clocks the scan path. ShiftDR is derived from the similarly named state in the TAP controller of Figure 11.19. ClockDR is asserted when the TAP controller is in state Capture-DR or Shift-DR. (Hence, of course, the boundary scan architecture is not truly synchronous!)

3. Capture mode. ShiftDR selects the IN input; data is clocked into the scan path register with ClockDR to take a snapshot of the system.

4. Update mode. After a capture or scan, data from the left flip-flop is sent to OUT by applying one clock edge to UpdateDR. Again, this clock signal comes from the TAP

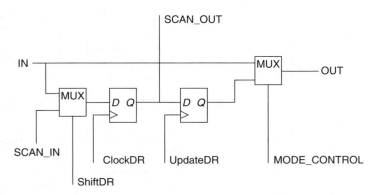

Figure 11.20   Boundary scan cell.

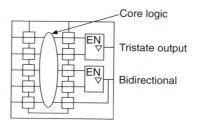

Figure 11.21   Logic outside boundary scan path.

controller when it is in state Update-DR. MODE_CONTROL is set as appropriate according to the instruction held in the instruction register (see below).

For normal input and output pins, the boundary scan cells are the only logic between the core and the IC pins. The only cases where logic is permitted between the boundary scan cell and an external pin are shown in Figure 11.21.

A number of instructions may be loaded into the instruction register. These allow specific tests to be performed. Three of these tests are mandatory; the remaining tests are optional. Some of these tests are:

- EXTEST (mandatory). This instruction performs a test of the system, external to the core logic of particular devices. Data is sent from the output boundary scan cells of one device, through the pads and pins of that device, along the interconnect wiring, through the pins and pads of a second device and into the input boundary scan cells of that second device. Hence a complete test of the interconnect from one IC core to another is performed.

- SAMPLE/PRELOAD (mandatory). This instruction is executed before and after the EXTEST and INTEST instructions to set up pin outputs and to capture pin inputs.

- BYPASS (mandatory). This instruction selects the Bypass register to shorten the scan path.

- RUNBIST (optional). Runs a built-in self-test on a component.

- INTEST (optional). This instruction uses the boundary scan register to test the internal circuitry of an IC. Although such a test would normally be performed before a component is mounted on a PCB, it might be desirable to check that the process of soldering the component onto the board has not damaged it. Note that the internal logic is disconnected from the pins, so if pins have been connected together on the board, that will have no effect on the standard test.

- IDCODE, USERCODE (optional). These instructions return the identification of the device (and the user identification for a programmable logic device). The code is put into the scan path.

- CONFIGURE (optional). An SRAM-based FPGA needs to be configured each time power is applied. The configuration of the FPGA is held in registers. These registers can be linked to the TAP interface. This clearly saves pins as the configuration and test interfaces are shared.

The MODE_CONTROL signal of Figure 11.20 is set to select the flip-flop output when instructions EXTEST, INTEST and RUNBIST are loaded in the instruction register. Otherwise the IN input is selected.

Testing a board with boundary scan components is similar in many ways to testing a component with a scan path. First, the boundary scan circuitry itself must be tested for faults such as a broken scan path or a TAP failure. Then interconnect and other tests can be performed. The boundary scan path allows nodes to be controlled from one point in the scan path and observed at another point. Test patterns for the interconnect (and for non-boundary scan-compliant components) have to be derived in much the same way that tests for logic are determined. These tests and the appropriate instructions have to be loaded into the registers of boundary scan components in the correct order. This process is clearly complex to set up and really has to be automated.

An example of how boundary scan might be included on an IC is shown in Figure 11.22. The basic circuit has two D-type flip-flops with a clock and reset. The D, Q, clock and reset pins have boundary scan cells included as shown. A TAP controller and instruction and bypass registers are included, together with the four extra pins.

In order that boundary scan-compliant components from different manufacturers may be used together, a standard description language – Boundary Scan Description Language (BSDL) – has been defined. This is a subset of VHDL. A BSDL description of the IC of Figure 11.22 is shown below. It is not appropriate here to describe every detail. The standard (1149.1-2001) defines a number of attributes. The BSDL description of the device consists of an **entity** with a **generic** and a **port**. Note that the VDD and GND pins are included in the port, with the mode **linkage**. The attributes are included in the **package** STD_1149_1_2001, which is referenced in the middle of the entity description so that its scope covers only the succeeding part of the entity description. (As the standard acknowledges, this use clause may need to be amended to include the library into which the package has been compiled.) The attributes and constant in the description define the pin mapping, the TAP signals present, the instructions that are implemented and the bit patterns that implement them, and the structure of the boundary scan register, including which type of boundary scan cell is used for each pin.

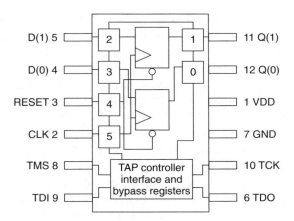

Figure 11.22   IC with boundary scan.

```
entity dff_2 is
  generic(PHYSICAL_PIN_MAP : string := "UNDEFINED");
  port (CLK : in BIT; RESET : in BIT;
        Q : out BIT_VECTOR(1 to 2);
        D : in BIT_VECTOR(1 to 2);
        GND, VDD : linkage BIT;
        TDO : out bit; TMS, TDI, TCK : in BIT);
  use STD_1149_1_2001.all;
  attribute COMPONENT_CONFORMANCE of dff_2 : entity is
    "STD_1149_1_2001";
  attribute PIN_MAP of dff_2 : entity is PHYSICAL_PIN_MAP;
  constant DIL_PACKAGE : PIN_MAP_STRING :=
    "CLK:2, RESET:3, Q:(12,11), D:(4,5), GND:7, VDD:1," &
    "TDO:6, TMS:8, TDI:9, TCK:10";
  attribute TAP_SCAN_IN of TDI : signal is TRUE;
  attribute TAP_SCAN_MODE of TMS : signal is TRUE;
  attribute TAP_SCAN_OUT of TDO : signal is TRUE;
  attribute TAP_SCAN_CLOCK of TCK : signal is (20.0E6,
        BOTH);
  attribute INSTRUCTION_LENGTH of dff_2 : entity is 2;
  attribute INSTRUCTION_OPCODE of dff_2 : entity is
    "Bypass (11), Extest (00), Sample (01)";
  attribute INSTRUCTION_CAPTURE of dff_2 : entity is "01";
  attribute BOUNDARY_LENGTH of dff_2 : entity is 6;
  attribute BOUNDARY_REGISTER of dff_2 : entity is
--num cell port function safe
    "5 (BC_1, CLK, input, X)," &
    "4 (BC_1, RESET, input, X)," &
    "3 (BC_1, D(1), input, X)," &
    "2 (BC_1, D(2), input, X)," &
    "1 (BC_1, Q(2), output2, X)," &
    "0 (BC_1, Q(1), output2, X)";
end dff_2;
```

The costs of implementing boundary scan on an integrated circuit include the cost of a boundary scan cell for each pin, the TAP controller, the one-bit bypass register, the instruction register and four extra pins. There will be extra wiring on the PCB.

On the other hand there can be significant benefits. The fault coverage of a PCB can be close to 100%. Boundary scan is easy to implement on a PCB requiring four pins on an edge connector. Specialist, expensive test equipment, such as a bed-of-nails tester, is not needed. Indeed, it is possible to implement a boundary scan tester using little more than a standard personal computer or workstation. Tests can be performed on ICs after they have been mounted on the PCB, so field testing is easy. Because the test circuitry is independent of normal system functions, it is possible to monitor the inputs and outputs of ICs in normal operation, thus providing debugging functions.

There are an increasing number of ICs with boundary scan compliance, e.g. Intel Pentium, Motorola 68040 and Xilinx programmable logic.

Summary

The testability of a circuit can be improved by modifying the circuit design. The simplest modifications include providing asynchronous resets to every register and avoiding redundant and other uncontrollable logic. SISO separates the sequential from the combinational logic, reducing test generation to a purely combinational circuit problem. Built-in self-test (BIST) can reduce manufacturing costs by putting much of the test circuitry on the chip. Boundary scan uses the SISO principle to allow complex PCBs to be tested. These various techniques can be combined.

Further reading

The books by Abramovici, Breuer and Friedman, by Miczo and by Wilkins all describe design for test methods. Boundary scan is now incorporated into many FPGAs, and the TAP interface is used to configure the internal logic. Details are on the manufacturers' websites. Agilent offer an online BSDL checking facility.

Exercises

11.1 Explain what is meant by initialization. Why is it necessary to initialize a circuit for test purposes even if it is not necessary in its system function?

11.2 What are the problems that the scan-in, scan-out (SISO) method is intended to overcome? Explain the principles of the SISO method, and identify the benefits and costs involved.

11.3 A certain integrated circuit contains 50 D-type flip-flops. Assuming that all states are reachable, and that it may be clocked at 1 MHz, what is the minimum time needed for an exhaustive test? If the same integrated circuit is designed with a full scan path and if all the combinational logic may be fully tested with 200 test vectors, estimate the time now required to complete a full test.

11.4 Show that the circuit of Figure 11.23 is a suitable test generator for an n-input NAND gate. Hence suggest a suitable BIST structure for each of the NAND planes in a PLA.

11.5 Figure 11.24 shows the structure of a simple CPU (reproduced from Chapter 7). There is a single bus, 8 bits wide. 'PC', 'IR', 'ACC', 'MDR' and 'MAR' are 8-bit registers. 'Sequencer' is a state machine with inputs from the 'IR' block and from other points in the system and with outputs that control the operation of the 'ALU' and that determine which register drives the bus.

The CPU design is to be modified to include a self-test facility. This self-test will not require the use of any external signals or data other than the clock and

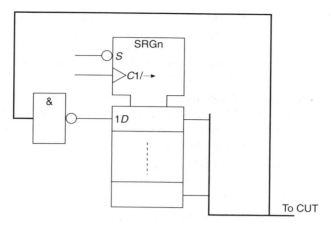

Figure 11.23   Circuit for Exercise 11.4.

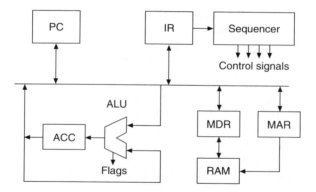

Figure 11.24   CPU datapath for Exercise 11.5.

will generate a simple pass/fail indication. The self-test should require as little additional hardware as possible.

(a) Describe the modifications you would make to the hardware to allow a self-test to be performed.

(b) Describe the strategy to be used to test the system, excluding the 'Sequencer'. Does testing the 'ALU' present any particular difficulties?

11.6   What are the main hardware components of the IEEE 1149.1 boundary scan test architecture?

11.7   Figure 11.19 shows the state transition diagram of the boundary scan TAP controller. Assuming that the instruction for an EXTEST is 10 for a particular IC, what sequence of inputs needs to be applied to the TAP of that IC to load the pattern 1010 into the first four stages of the boundary scan register of the IC and to run an EXTEST? (Note that the least significant bits should be loaded first.)

11.8 If the outputs of four boundary scan register stages are connected to the inputs of four similar register stages in a second IC, show, in principle, how the test sequence from Exercise 11.7 can be extended to capture the responses of the interconnect. What assumptions have you made about the connection of the test structures on the two ICs?

11.9 A particular integrated circuit has 2000 flip-flops and 5000 other gates. The package has 52 pins, including power, ground, clock and reset. All the buses are 16 bits wide. A new version of the circuit is to be built. Before redesigning the circuit, the manufacturer would like an estimate of the costs of including:

(a) one or more scan paths to cover all of the flip-flops;
(b) boundary scan to IEEE 1149.1 standard;
(c) built-in self-test.

The estimates should be in terms of extra components and pins and should consider each of the three features individually, together with any savings that may be made by including two or more features.

11.10 Write a synthesizable VHDL model of the IEEE 1149.1 TAP controller. The following outputs should be asserted:

Signal	State(s)
UpdateDR	Update-DR
ClockDR	Capture-DR
Shift-DR	ShiftDR
Shift-DR	UpdateIR
Update-IR	ClockIR
Capture-IR	Shift-IR
ShiftIR	Shift-IR

11.11 Modify the VHDL model of the LFSR from Chapter 6 to implement an n-stage MISR. Hence, write a model of an n-bit BILBO register.

Chapter 12

Asynchronous sequential design

The sequential circuits described in Chapters 5, 6 and 7 are synchronous. A clock is used to ensure that all operations occur at the same instant. This avoids the problems of hazards, because such transient effects can be assumed to have died away before the next clock edge. Therefore irredundant logic can be used, which then makes the combinational parts of the circuits fully testable, at least in theory. The flip-flops used in synchronous design are, however, asynchronous internally. In this chapter we will consider the design of asynchronous elements and use a VHDL simulator to illustrate the difficulties of asynchronous design.

12.1 Asynchronous circuits

Throughout this book, the emphasis has been on the design of *synchronous* sequential circuits. State information or other data has been loaded into flip-flops at a clock edge. Asynchronous inputs to flip-flops have been used, but *only* for initialization. A common mistake in digital design is to use these asynchronous inputs for purposes other than initialization. This mistake is made either because of inexperience or because of a desire to simplify the logic in some way. Almost inevitably, however, circuits designed in such a manner will cause problems, by malfunctioning or because subsequent modification or transfer to a new technology will cause the assumptions made in the design to become invalid.

Synchronous sequential design is almost overwhelmingly preferred and practised because it is easier to get right than asynchronous design. Simply connecting logic to the asynchronous inputs of flip-flops is almost always wrong. Structured design techniques exist for asynchronous design and this chapter will describe the design process and its pitfalls. It should be noted, however, that we are primarily concerned with the design of circuits comprising a few gates. It is possible to design entirely asynchronous systems, but such methodologies are still the subject of research. Nevertheless, as clock speeds increase, some of the complex timing issues described here will become relevant. It is increasingly difficult to ensure that a clock edge arrives at every flip-flop in a system at *exactly* the same instant. Systems may consist of synchronous islands that communicate asynchronously. To ensure such communications are as reliable as possible, specialized interface circuits will need to be designed, using the techniques described in this chapter.

Although, as noted above, this book has been concerned with synchronous systems, reference was made to the synthesis of asynchronous elements in Chapter 9. At present, synthesis tools are intended for the design of synchronous systems, normally with a single clock. This is particularly true of synthesis tools intended for FPGA design. The concurrent VHDL construct

```
q <= d when c = '1' else q;
```

would be synthesized to an asynchronous sequential circuit structure. Similarly, the sequential block

```
process (d, c) is
begin
  if c = '1' then
    q <= d;
  end if;
end process;
```

would also be synthesized to an asynchronous latch. In both cases, q explicitly holds on to its value unless c is asserted. It might be thought that the circuit structures created by a synthesis tool for the two cases would be identical. In general, this is not so. The first case is exactly the same as writing

```
q <= (d and c) or (q and not c);
```

Hence, a synthesis tool would create an inverter, two AND gates and an OR gate (or an inverter and three NAND gates). On the other hand, a 1076.6 compliant synthesis tool would infer the existence of a latch from the incomplete **if** statement of the second case, and use a latch from a library (while also issuing a warning message, in case the incomplete **if** statement were a coding error). The latch created by Boolean minimization and the library latch are not the same. Indeed, the RTL synthesis standard, IEEE 1076.6, explicitly forbids the use of concurrent assignments of the form shown, while permitting the use of incomplete **if** and **case** statements.

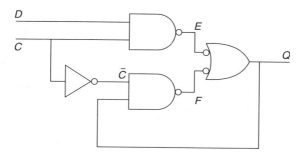

Figure 12.1 Basic D latch.

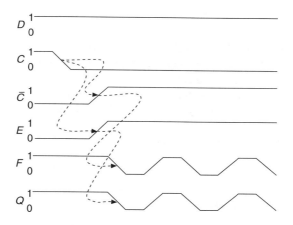

Figure 12.2 Timing diagram for circuit of Figure 12.1.

To see why, assume that the circuit has been implemented directly, as shown in Figure 12.1. This circuit should be compared with that of Figure 2.13. Indeed, the following analysis is comparable with that of Section 2.4. Let us assume that each gate, including the inverter, has a delay of one unit of time, e.g. 1 ns. Initially, Q, D and C are at logic 1. C then changes to 0. From the analysis of Section 2.4, we know that this circuit contains a potential hazard. When we draw a timing diagram for this circuit, as shown in Figure 12.2, this hazard appears at Q. This hazard is propagated back to F, which causes Q to change *ad infinitum*. Hence the circuit oscillates. The causality between F and Q is not shown in Figure 12.2, for clarity. This kind of behaviour is obviously extremely undesirable in a sequential circuit. Although the assumption of a unit delay in each gate may be unrealistic, it can easily be demonstrated, by means of a VHDL simulation, that a hazard and hence oscillatory behaviour will occur, irrespective of the exact delays in each gate.

We should, at this point, include a very clear warning. Although we will use VHDL in this chapter to model and to simulate the behaviour of asynchronous circuits, these simulations are intended to demonstrate that problems *may* exist. It is extremely difficult to accurately predict by simulation *exactly* how a circuit will

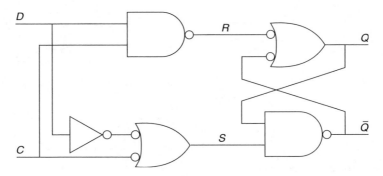

Figure 12.3   D latch with hazard removed.

behave, particularly when illegal combinations of inputs are applied. The spurious effects result from voltage and current changes within electronic devices, not transitions between logic values.

The solution to the problem of oscillatory behaviour is, as stated in Section 2.4, to include redundant logic by way of an additional gate. Thus,

$$Q^+ = D.C + Q.\overline{C} + D.Q$$

or

$$Q^+ = \overline{\overline{D.C}.\overline{Q.\overline{C}}.\overline{D.Q}}$$

where Q^+ represents the 'next' value of Q. The redundant gate, $\overline{D.Q}$, has a 0 output while D is 1, therefore Q is held at 1.

The expression for Q^+ can be rearranged:

$$Q^+ = D.C + Q.(\overline{C} + D)$$

Hence the circuit of Figure 12.3 can be constructed. This would not and could not be generated by optimizing logic equations, but instead would exist in a library. It is this circuit that would be called from the library by a synthesis tool when an incomplete **if** statement was encountered.

12.2   Analysis of asynchronous circuits

12.2.1   Informal analysis

The operation of the D latch of Figure 12.3 is relatively straightforward. The key is the operation of the cross-coupled NAND gates. Two NAND (or NOR) gates connected in this way form an RS latch with the truth table given below. (An RS

latch built from NOR gates has a similar truth table, but with the polarities of R and
S reversed.)

R	S	Q^+	$\overline{Q}^+$
0	0	1	1
0	1	1	0
1	0	0	1
1	1	Q	$\overline{Q}$

The input $R = S = 0$ is normally considered illegal, because it forces the outputs to be
the same, contradicting the expected behaviour of a latch.

The D latch of Figure 12.3 contains an RS latch, in which R and S are controlled by
two further NAND gates. When C is at logic 0, R and S are at 1; therefore the latch
holds whatever value was previously written to it. When C is 1, S takes the value of
D and R takes the value of $\overline{D}$. From the truth table above we can see that Q therefore
takes the value of D. We can further note that the signal paths from D to the outputs are
unequal, because of the inverter. It is therefore reasonable to assume that if D and C
were to change at the same time, the behaviour of the latch would be unpredictable.

Figure 12.4 shows the circuit of a positive edge-triggered D flip-flop. We will attempt
to analyze this circuit informally, but this analysis is intended to show that a formal
method is needed. Let us first deal with the 'asynchronous' set and reset.[1] If S is 0 and
R is 1, Q is forced to 1 and $\overline{Q}$ is forced to 0, according to the truth table above. Similarly,
if S is 1 and R is 0, Q is forced to 0 and $\overline{Q}$ is forced to 1. Under normal synchronous
operation, S and R are both held at 1, and therefore can be ignored in the following

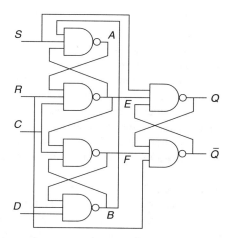

Figure 12.4   Positive edge-triggered D flip-flop.

---

[1]At this level, all the inputs are asynchronous, of course. Synchronous design works because we follow certain conventions about the use of inputs, not because particular inputs are special.

analysis. Note, however, that if both S and R are held at 0, both Q and $\overline{Q}$ go to 1; hence this condition is usually deemed to be illegal.

Let us consider the effects of changes at the D and C inputs while $R = S = 1$. If C is at 0, then both E and F are at 1 and therefore Q and $\overline{Q}$ are held. If D is at 0, internal nodes A and B are at 0 and 1, respectively. If D is at 1, A is 1 and B is 0. Therefore D can change while the clock is low, causing A and B to change, but further changes, to E and F, are blocked by the clock being low.

When the clock changes from 0 to 1, *either* D is 0, and hence A is 0 and B is 1, which force E to 1 and F to 0 and therefore Q to 0 and $\overline{Q}$ to 1, *or* D is 1, A is 1, B is 0 and therefore E is 0, F is 1, Q is 1 and $\overline{Q}$ is 0. Therefore when the clock changes, it is assumed that A and B are stable. Hence, there is a *setup time* in which any change in D must have propagated to A, before the clock edge.

While the clock is 1, D can again change without affecting the outputs. Two conditions are possible: (a) D was 0 at the clock edge, and hence A is 0, B is 1, E is 1 and F is 0. If D changes to 1, there will be no change to B, because F is 0 and hence B is always 1; or (b) D was 1 at the clock edge, thus A is 1, B is 0, E is 0 and F is 1. If D changes to 0, B changes from 0 to 1, but as E is 0, this change is not propagated to A. Therefore, again, the output is unaffected. The falling clock edge forces both E and F to 1 again.

It is apparent that this descriptive, intuitive form of analysis is not sufficient to adequately describe the behaviour of even relatively small asynchronous circuits. Moreover, it would be impossible to design circuits in such a manner. It is possible to use a VHDL simulator to verify the behaviour of such circuits, but we need a formal analysis technique.

12.2.2   Formal analysis

Before proceeding with the formal analysis of both the D latch and the edge-triggered D flip-flop, we need to state a basic assumption. The *fundamental mode* restriction states that only one input to an asynchronous circuit may change at a time. The effects of an input change must have propagated through the circuit and the circuit must be stable before another input change can occur. The need for this restriction can be seen from the two circuits already considered. If D changes at almost the same time as the clock, unequal delay paths mean that internal nodes are not at expected, consistent values and unpredictable behaviour may result. In the worst case the output of a latch or flip-flop may be in an intermediate *metastable* state, that is neither 0 nor 1. We will return to metastability later.

In order to perform a formal analysis, we have to break any feedback loops in the circuit. Of course, we don't actually change the circuit, but for the purposes of the analysis we pretend that all the gate delays in the circuit are concentrated in one or more *virtual buffers* in the feedback loops. The gates are therefore assumed to have zero delays. The D latch is redrawn in Figure 12.5. Note that there is only one feedback loop in this circuit, although at first glance the cross-coupled NAND gate pair might appear to have two feedback loops. If the one feedback loop were really broken, the circuit would be purely combinational, which is sufficient. In Figure 12.5, the input to the virtual buffer is labelled as Y^+, while the

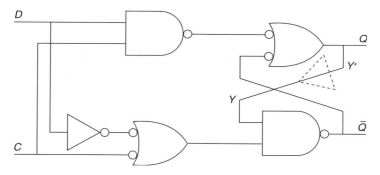

Figure 12.5   D latch with virtual buffer.

		DC		
Y	00	01	11	10
0	0	0	1	0
1	1	0	1	1
			Y⁺	

Figure 12.6   Transition table for D latch.

output is labelled as Y. Y is the *state variable* of the system. This is analogous to the state variable in a synchronous system. Y^+ is the next state. The system is *stable* when Y^+ is equal to Y. In reality, of course, Y^+ and Y are two ends of a piece of wire and must have the same value, but, to repeat, for the purpose of analysis we pretend that they are separated by a buffer having the aggregate delay of the system. Note that we separate the state variable from the output, although in this case, Q and Y^+ are identical.

We can write the state and output equations for the latch as:

$$Y^+ = D.C + Y.\overline{C} + D.Y$$
$$Q = D.C + Y.\overline{C} + D.Y$$
$$\overline{Q} = \overline{D}.C + \overline{Y}$$

From this we can now write a *transition table* for the state variable, as shown in Figure 12.6.

A *state table* replaces the Boolean state variables with abstract states. In the state table of Figure 12.7 the stable states are circled. A state is stable when the next state is equal to the current value. The state table can also include the outputs (*state and output table*), as shown in Figure 12.7. Notice that there is an unstable state that has both outputs the same.

Using the state and output table, we can trace the change of states when an input changes. Starting from the top left corner of the table, with the current state as K and the two inputs at 0, let D change to 1. From Figure 12.8, it can be seen that the state and output remain unchanged. If C then changes to 1, the system moves into an unstable state. The system now has to move to the stable state at L, with D and C both equal to 1.

	DC			
S	00	01	11	10
K	Ⓚ,01	Ⓚ,01	L,11	Ⓚ,01
L	Ⓛ,10	K,01	Ⓛ,10	Ⓛ,10

$S^+, Q\bar{Q}$

Figure 12.7   State table for D latch.

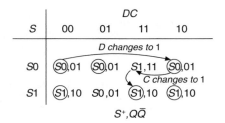

	DC			
S	00	01	11	10
		D changes to 1		
S0	Ⓢ0,01	Ⓢ0,01	S1,11	Ⓢ0,01
				C changes to 1
S1	Ⓢ1,10	S0,01	Ⓢ1,10	Ⓢ1,10

$S^+, Q\bar{Q}$

Figure 12.8   Transitions in state table.

Note that the state transition *must* be a vertical move on the state transition diagram. This is in order to comply with the fundamental mode restriction – anything other than a vertical move implies a change in an input value, which would therefore be occurring before the system was stable. It can be seen that the latch behaves as we would expect a *D* latch to behave. If *D* is changed from 0 to 1, followed by *C* changing from 0 to 1, we would expect *Q* to change from 0 to 1, and it can be seen from Figure 12.8 that this is what happens.

12.3 Design of asynchronous sequential circuits

In essence, the design procedure for asynchronous sequential circuits is the reverse of the analysis process. An abstract state table has to be derived, then a state assignment is performed, and finally state and output equations are generated. As will be seen, however, there are a number of pitfalls along the way, making asynchronous design much harder than synchronous design. To illustrate the procedure, we will perform the design of a simple circuit, and show, both theoretically and by simulation, the kinds of errors that can be made.

Let us design an asynchronous circuit to meet the following specification: the circuit has two inputs, *Ip* and *Enable*, and an output, *Q*. If *Enable* is high, a rising edge on *Ip* causes *Q* to go high. *Q* stays high until *Enable* goes low. While *Enable* is low, *Q* is low.

It can be see from this specification that there are eight possible combinations of inputs and outputs, but that two combinations cannot occur: if *Enable* is low, *Q* cannot be high. This leaves six states to the system, as shown in Table 12.1.

Table 12.1	States of example asynchronous system.		
State	Ip	Enable	Q
a	0	0	0
b	0	1	0
c	1	0	0
d	1	1	0
e	0	1	1
f	1	1	1

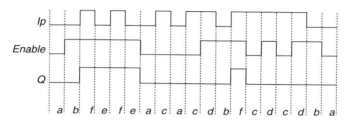

Figure 12.9   States in design example.

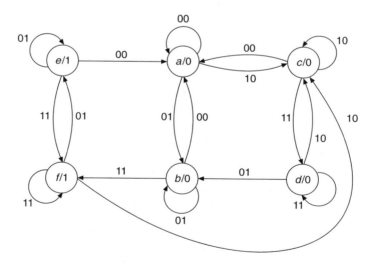

Figure 12.10   State transition diagram for design example.

The first task is to work out all the possible state transitions. One way to do this is to sketch waveforms and to mark the states as shown in Figure 12.9. From this a state transition diagram can be constructed (Figure 12.10). This state diagram can also be expressed as the *primitive flow table* of Figure 12.11. A primitive flow table has one state per row. Because of the fundamental mode restriction, only state transitions that are reachable

Ip Enable

S	00	01	11	10	Q
a	(a)	b	–	c	0
b	a	(b)	f	–	0
c	a	–	d	(c)	0
d	–	b	(d)	c	0
e	a	(e)	f	–	1
f	–	e	(f)	c	1

S^+

Figure 12.11   Primitive flow table.

Ip Enable

S	00	01	11	10	Q
A	(A)	(A)	E	C	0
C	A	A	(C)	(C)	0
E	A	(E)	(E)	C	1

S^+

Figure 12.12   State and output table.

from a stable state with one input change are marked. State transitions that would require two or more simultaneous input changes are marked as 'don't cares'. The outputs are shown for the stable state and all transitions out of the state. It is also possible to assume that the outputs only apply to the stable states and that the outputs during all transitions are 'don't cares'.

There are more states in this primitive flow table than are needed. In Chapter 5, it was shown that states can be merged if they are equivalent. In this example, there are 'don't care' conditions. We now speak of states being *compatible* if their next states and outputs are the same or 'don't care'. There is an important difference between equivalence and compatibility. It can be seen that states *a* and *b* are compatible and that states *a* and *c* are compatible. States *b* and *c* are, however, not compatible. If *a* and *b* were *equivalent* and *a* and *c* were also equivalent, *b* and *c* would be equivalent by definition.

Here, states *a* and *b* are compatible and may be merged into state *A*, say. When compatible states are merged, 'don't cares' are replaced by defined states or outputs (if they exist). Similarly, states *c* and *d* may be merged into *C*, and *e* and *f* may be merged into *E*. The resulting state and output table is shown in Figure 12.12.

At this point, considerable care is needed in making an appropriate state assignment. We will first demonstrate how *not* to perform a state assignment. We can show,

Y_1Y_0	00	01	11	10	Q
	Ip Enable				
00	(00)	(00)	11	01	0
01	00	00	(01)	(01)	0
11	00	(11)	(11)	01	1
	$Y_1^+Y_0^+$				

Figure 12.13 Transition table.

using a VHDL simulation, that a poor state assignment can easily result in a malfunctioning circuit. To encode three states requires two state variables, as described in Chapter 5. There are 24 possible state assignments. As with a synchronous system, there is no way to tell, in advance, which state assignment is 'best'. Therefore, let us arbitrarily assign 00 to *A*, 01 to *C* and 11 to *E*. This gives the transition table shown in Figure 12.13. The state 10 is not used, so in deriving next state expressions, the entries corresponding to 10 are 'don't cares'. Hazard-free next state and output equations can be found using K-maps:

$$Y_1^+ = Y_1.Enable + Ip.Enable.\overline{Y_0}$$
$$Y_0^+ = Ip + Y_1.Enable$$
$$Q = Y_1$$

A VHDL model of this circuit is as follows. The next state expressions have been given arbitrary delays. It is left as an exercise for the reader to write a suitable testbench.

```
library IEEE;
use IEEE.std_logic_1164.all;

entity Asynch_Ex is
  port (ip, enable : in std_logic;
        q : out std_logic);
end entity Asynch_Ex;

architecture Version1 of Asynch_Ex is
  signal y1, y0 : std_logic;
begin
  y1 <= (y1 and enable) or (ip and enable and (not y0))
          after 3 NS;
  y0 <= ip or (y1 and enable) after 2 NS;
  q <= y1;
end architecture Version1;
```

If Y_1 and Y_0 are both 0 and *Ip* and *Enable* are 0 and 1, respectively, Q is 0. Now, let *Ip* change to 1. We would expect to move horizontally into an unstable state and then to move vertically to the stable state $Y_1Y_0 = 11$. In fact, the VHDL simulation shows that

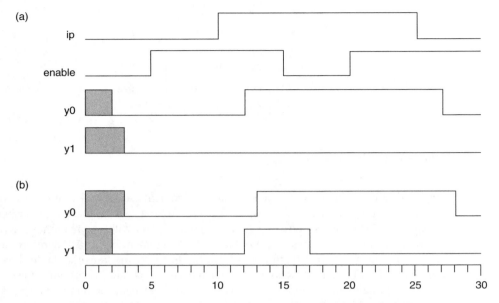

Figure 12.14 Simulation of asynchronous circuit example: (a) with race; (b) without race.

Figure 12.15 Transition table with critical race.

the circuit goes to $Y_1Y_0 = 01$ (Figure 12.14(a)). If the delays are reversed, however, the circuit works as expected (Figure 12.14(b)):

```
y1 <= (y1 and enable) or (ip and enable and (not y0))
        after 2 NS;
y0 <= ip or (y1 and enable) after 3 NS;
```

Why is the circuit sensitive to these delays? We have accounted for hazards in the Boolean minimization, so they are not the problem. Let us consider the transition table, including the unused state, with the values for the unused state as implied by the minimized equations, as shown in Figure 12.15.

In the first case, Y_1 changes first; therefore the circuit changes to the unstable state 10, at which point Y_0 changes and the circuit finishes in the correct state. In the second case, Y_0 changes first and the circuit moves to the stable state 01, *and stays there*! In other words, the order in which the state variables change can affect the final state of the circuit. The situation in which two or more state variables change as a result of one input change is known as a *race*. If the final state depends on the exact order of the state variable changes, that is known as a *critical race*. There is a potentially even more disastrous situation. If the don't cares in the K-maps produced from the transition table of Figure 12.13 were forced to be 0 (which results in non-minimal next state expressions, but is otherwise perfectly legitimate), the next state equations become:

$$Y_1^+ = Y_1.Y_0.Enable + Ip.Enable.\overline{Y_1}.\overline{Y_0}$$
$$Y_0^+ = Ip.\overline{Y_1} + Ip.Y_0 + Y_1.Y_0.Enable$$

When the VHDL model shown below is simulated, the circuit oscillates, as shown in Figure 12.16.

```
y1 <= (y1 and y0 and enable) or
      (ip and enable and (not y1) and (not y0)) after 2 NS;
y0 <= (ip and (not y1)) or (ip and y0) or
      (y1 and y0 and enable) after 3 NS;
```

Figure 12.17 shows the transition table. Y_1 changes to 1 before Y_0 can react, so the circuit moves to state 10. Y_1 is then forced back to 0, so the circuit oscillates between states 00 and 01. This is known as a *cycle*.

We clearly have to perform a state assignment that avoids both critical races and cycles. In this example, such an assignment is not possible with just three states. Therefore we have to introduce a fourth state. This state is unstable, but it ensures that only one state variable can change at a time. Figure 12.18 shows the modified state table, while Figure 12.19 shows a simplified state transition diagram, with the newly

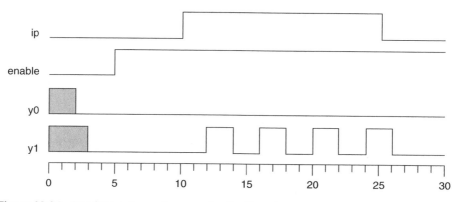

Figure 12.16   Simulation of asynchronous circuit with cycle.

	Ip Enable				
$Y_1 Y_0$	00	01	11	10	Q
00	(00)	(00)	11	01	0
01	00	00	(01)	(01)	0
11	00	(11)	(11)	01	1
10	00	00	00	00	1

$$Y_1{}^+ Y_0{}^+$$

Figure 12.17 Transition table with cycle.

	Ip Enable				
S	00	01	11	10	Q
A	(A)	(A)	G	C	0
C	A	A	(C)	(C)	0
E	G	(E)	(E)	C	1
G	A	–	E	–	–

$$S^+$$

Figure 12.18 Modified state table.

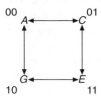

Figure 12.19 Simplified state transition diagram.

introduced state, G, and a suitable state assignment. Hence expressions for the state variables can be derived. In this case, the state variable expressions are:

$$Y_1{}^+ = Y_1.Y_0.\,\overline{Ip} + Y_1.Enable + Ip.Enable.\overline{Y_0}$$
$$Y_0{}^+ = Ip.\overline{Enable} + Ip.Y_0 + Y_1.Enable$$

We can simulate VHDL models of this circuit with either Y_1 or Y_0 changing first, and in both cases the circuit works correctly.

There is, however, one final potential problem. There are no possible redundant terms in this example, so we can be sure that all potential static hazards have been eliminated. In principle, therefore, the circuit can be built as shown in Figure 12.20.

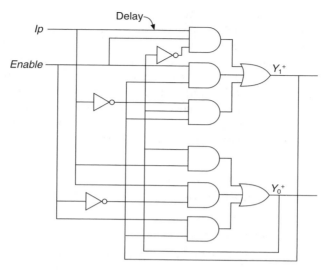

Figure 12.20 Circuit with essential hazard.

If, however, as a result of the particular technology used or the particular layout adopted, the input to the top AND gate is delayed with respect to the state variables, as shown, the circuit may still malfunction. This condition can be demonstrated again with a VHDL model:

```
islow <= ip after 5 NS;
y1 <= (y1 and y0 and (not ip)) or (islow and enable and
       (not y0)) or
       (enable and y1) after 2 NS;
y0 <= (ip and y0) or (ip and (not enable)) or
       (y1 and enable) after 3 NS;
```

The transition table of Figure 12.21 shows what happens if *Ip* changes from 1 to 0 from state 01 while *Enable* stays at 1. In theory this change should cause only transitions

$Y_1 Y_0$	00	01	11	10	Q
00	(00)	(00)	10	01	0
01	00	00	(01)	(01)	0
11	10	(11)	(11)	01	1
10	00	–	11	–	–

Ip Enable

$Y_1{}^+Y_0{}^+$

Figure 12.21 Transition table with essential hazard.

1*a* and 1*b* and the final state should be 00. In practice, because of the delay in *Ip*, the circuit then follows the other transitions shown, 2*a*, 2*b*, 3*a* and 3*b*, to finish in state 11. This is known as an *essential hazard*, so-called because it is part of the essence of the circuit. Potential essential hazards can be identified from the transition table if a single input change results in a different final state than if the input changes three times. The only way to avoid *essential hazards* is to ensure that the state variables cannot be fed back round the circuit before the input transitions. This can be achieved by careful layout or possibly by deliberately introducing delays into the state variables.

In summary, therefore, the design of an asynchronous sequential circuit has the following steps:

1. State the design specifications.
2. Derive a primitive flow table.
3. Minimize the flow table.
4. Make a race-free state assignment.
5. Obtain the transition table and output map.
6. Obtain hazard-free state equations.
7. Check for essential hazards.

(12.4) Asynchronous state machines

In the design flow, above, the first step is to derive the design specifications. In many ways this is the hardest part of the task. Moreover, if we get that wrong, everything that follows is also by definition wrong. By the nature of the design process, it is almost impossible to patch a mistake – the entire process has to be repeated. Therefore, it would be very desirable to ensure that the design has been specified correctly. One way to do this is to use simulation again.

The state transition diagram of Figure 12.10 is essentially the same as the state diagram of Figure 5.9 or that of Figure 11.19. One figure represents an asynchronous system and two represent a synchronous system. This difference is not, however, apparent from the diagrams. We advocated the use of ASM charts for the design of synchronous systems, but we could have used state diagrams. We know that an ASM chart or a state diagram has an equivalent VHDL description. By the same argument, we can represent an asynchronous state machine in VHDL. Instead of a set of registers synchronized to a clock, we would have a virtual buffer, in which the state variable is updated. Let us therefore write a VHDL description of the state machine of Figure 12.10.

```
library IEEE;
use IEEE.std_logic_1164.all;

entity Asynch_Ex is
  port (ip, enable : in std_logic;
        q : out std_logic);
end entity Asynch_Ex;
```

```
architecture state_mc of Asynch_Ex is
  type state is (a, b, c, d, e, f);
  signal present_state, next_state : state;
begin
  -- virtual buffer
  present_state <= next_state after 1 NS;
  com: process (ip, enable, present_state) is
  begin
  q <= '0';
  case present_state is
    when a =>
      if ip = '0' and enable = '1' then
        next_state <= b;
      elsif ip = '1' and enable = '0' then
        next_state <= c;
      else next_state <= a;
      end if;
  -- other states are written the same way
    when f =>
      q <= '1';
      if ip = '0' and enable = '1' then
        next_state <= e;
      elsif ip = '1' and enable = '0' then
        next_state <= c;
      else next_state <= f;
      end if;
    end case;
  end process com;
end architecture state_mc;
```

The virtual buffer has a delay of 1 ns. For this type of model to work, there must be a finite delay – a zero delay would cause the process to loop infinitely at time 0. For reasons of space, the entire state machine is not shown; the other states may be written in the same way. The don't cares have been assumed to cause the state machine to stay in the same state. As these represent violations of the fundamental mode, this is valid. It is possible to check for fundamental mode violations by including an assert statement in the process:

```
assert ip'EVENT xor enable'EVENT xor present_state'EVENT
  report "Fundamental mode violation"
  severity NOTE;
```

This is not strictly correct as three simultaneous events would not trigger the assertion (see Exercise 12.11). With a suitable testbench, we can use this VHDL model to reproduce Figure 12.9. Notice that the initial values of the state variables will be the leftmost entry in the state definition – a.

We can also repeat the exercise after state minimization.

```
architecture reduced of Asynch_Ex is
  type state is (a, c, e);
  signal present_state, next_state : state;
begin
  -- virtual buffer
  present_state <= next_state after 1 NS;
  com: process (ip, enable, present_state) is
  begin
  q <= '0';
  case present_state is
    when a =>
      if ip = '1' and enable = '1' then
        next_state <= e;
      elsif ip = '1' and enable = '0' then
        next_state <= c;
      else next_state <= a;
      end if;
    when c =>
      if ip = '0' then
        next_state <= a;
      else next_state <= c;
      end if;
    when e =>
      q <= '1';
      if ip = '0' and enable = '0' then
        next_state <= a;
      elsif ip = '1' and enable = '0' then
        next_state <= c;
      else next_state <= e;
      end if;
    end case;
  end process com;
end architecture reduced;
```

Again, this can be verified by simulation. Indeed, this is one way to check that the state minimization has been done correctly.

As a second example, consider the following. We wish to design a phase detector with two outputs: *qA* and *qB*. There are also two inputs: *inA* and *inB*. Let us assume both outputs start high. When *inA* goes high, *qA* goes low and stays low until *inB* goes high. Similarly, if *inB* goes low first, *qB* goes low until *inA* goes high. This sounds very simple! We will model the phase detector as an asynchronous state machine. It is left as an exercise for the reader to derive the VHDL model, below, to implement this specification. You can further test your understanding of asynchronous design by taking this design through to gate level.

```vhdl
library IEEE;
use IEEE.std_logic_1164.all;

entity phase_detector is
  port (inA, inB : in std_logic;
        qA, qB : out std_logic);
end entity phase_detector;

architecture asynch_sm of phase_detector is
  type state_type is (A, B, C, D, E, F, G, H);
  signal present_state, next_state : state_type;
begin
  present_state <= next_state after 1 NS;
  process (inA, inB, present_state) is
  begin
    next_state <= present_state; -- default
    qA <= '1';
    qB <= '1';
    case present_state is
      when A =>
        if inA = '0' and inB = '1' then
          next_state <= E;
        elsif inA = '1' and inB = '0' then
          next_state <= B;
        end if;
      when B =>
        qA <= '0';
        if inA = '0' and inB = '1' then
          next_state <= D;
        elsif inA = '1' and inB = '1' then
          next_state <= C;
        end if;
      when C =>
        if inA = '0' and inB = '1' then
          next_state <= D;
        elsif inA = '1' and inB = '0' then
          next_state <= F;
        end if;
      when D =>
        if inA = '0' and inB = '0' then
          next_state <= A;
        elsif inA = '1' and inB = '1' then
          next_state <= H;
        end if;
      when E =>
        qB <= '0';
        if inA = '1' and inB = '1' then
```

```
              next_state <= C;
          elsif inA = '1' and inB = '0' then
              next_state <= F;
          end if;
        when F =>
          if inA = '0' and inB = '0' then
              next_state <= A;
          elsif inA = '1' and inB = '1' then
              next_state <= G;
          end if;
        when G =>
          qB <= '0';
          if inA = '0' and inB = '0' then
              next_state <= E;
          elsif inA = '0' and inB = '1' then
              next_state <= D;
          end if;
        when H =>
          qA <= '0';
          if inA = '0' and inB = '0' then
              next_state <= B;
          elsif inA = '1' and inB = '0' then
              next_state <= F;
          end if;
      end case;
    end process;
end architecture asynch_sm;
```

One final word of warning: do not try to synthesize these state machine models! In the light of the previous discussions, it should be obvious that you would generate hardware with races and hazards!

12.5 Setup and hold times and metastability

12.5.1 The fundamental mode restriction and synchronous circuits

The fundamental mode restriction requires that an input to an asynchronous circuit must not change until the circuit has become stable after a previous input change. Individual flip-flops are themselves asynchronous internally, but are used as synchronous building blocks. We do not, however, speak of the fundamental mode restriction when designing synchronous systems. Instead, we define setup and hold times.

Because of the gate delays in a circuit, the fundamental mode restriction *does not* mean that two inputs must not change at the exact same time. It means that the effect of one input change must have propagated through the circuit before the next input can change. To use the example of a D flip-flop, a change at the *D* input must have propagated through the flip-flop before an active clock edge may occur.

Similarly, the effect of the clock edge must have propagated through the circuit before the D input can change again. These two time intervals are known as the setup and hold times, respectively.

The setup and hold times of a latch or flip-flop depend on the propagation delays of its gates. These propagation delays depend, in turn, on parametric variations. So we can never know the exact setup and hold times of a given flip-flop. Furthermore, the timing of clock edges may be subject to *jitter* – the exact period of the clock may vary slightly. Therefore there has to be a margin of tolerance in estimating the setup and hold times. It should finally be noted that some of the effects of ignoring the fundamental mode restriction, or equivalently, violating setup and hold times, are not purely digital. In particular, metastability is effectively an analogue phenomenon.

Bearing all this in mind, it is possible to get some insight into the consequences of not observing the fundamental mode restriction by using a VHDL simulator. We could use trial and error to find the setup and hold times of a flip-flop; if the gate delays are specified, it is not difficult to calculate the various path lengths through the circuit. Here, however, we will use the random pulse stream generator from Chapter 6.

12.5.2   Random pulse generator

As seen in Chapter 6, random pulses can be generated using the pseudo-random function in the VHDL Math Package (1076.2). From a given pair of seeds, the same sequence will always be generated. This can be a good thing, because then an experiment is repeatable. On the other hand, this predictability may not be desirable. Elsewhere, it is common to use an integer such as the current time to randomize the seed. VHDL does not have this capability. The VHDL package below gets round this problem by saving the seeds between runs. Therefore with each simulation run, a seed will be read from a file to initialize the random number generator. At the same time, a new seed is generated and written back to the file. To repeat a simulation with the same seed, the seed file would need to be saved elsewhere and copied back (or the seed file is simply deleted to rerun the same simulation each time).

A **shared variable**, seed, is used to hold the state of the random number generator within a simulation run. In the 2002 VHDL standard, shared variables must be protected types. The package below defines a protected type. This is not compatible with the 1993 standard; Appendix C has a version of this package suitable for a simulator compliant with that standard. (The 1987 standard does not support shared variables at all.) The seed is initialized by calling the function init_seed, in which the value of the seed is read from a file, modified and written back to the same file, so that each time a simulation is run the seed will be different. This operation is done only once per simulation. There are two forms of the file_open function. Here the version with four parameters is used. The first parameter gives the status of the file operation, which allows us to take different actions depending on whether the file exists or not. A simple pseudo-random number generator uses seed to generate a real value between 0 and 1. Functions init_seed, get_seed and rand are **impure** because they read and modify the shared variable seed. The initial seed values and the modifying value are entirely arbitrary.

We could use the result of the random number generator, scaled and converted to a time, to cause random changes in a signal. It is more realistic to use a negative exponential function, as in Chapter 6. The mean time to the next event is specified, but instead of a uniform distribution half the event times will be between zero and the mean time, and half will be between the mean time and infinity. This model is commonly used in queuing theory. Two `negexp` functions are defined. The first generates a real number, given a real mean value. The second function uses the first `negexp` function, but takes a time as the mean value and returns the time to the next event in an integral number of nanoseconds. Note how the type and scale conversions are performed.

```
package random is
  impure function rand return REAL;
end package random;

library IEEE;
use IEEE.math_real.all;

package body random is
  type seed_pair is record
    seed1, seed2 : INTEGER;
  end record seed_pair;
  type pseed is protected
    impure function get_seed return seed_pair;
    impure function rand return REAL;
  end protected pseed;

  type pseed is protected body
    impure function init_seed return seed_pair is
      type natfile is file of INTEGER;
      file seedfile : natfile;
      variable status : FILE_OPEN_STATUS;
      variable seed1, seed2 : INTEGER;
    begin
      file_open(status, seedfile, "seed.dat", READ_MODE);
      assert status = OPEN_OK
        report "seed.dat not opened" severity NOTE;
      if (status = OPEN_OK) then
        read(seedfile, seed1);
        read(seedfile, seed2);
        file_close(seedfile);
      else
        seed1 := 56;
        seed2 := 42;
      end if;
      if seed1 < NATURAL'HIGH - 1000 then
        seed1 := seed1 + 1000;
      else
```

```vhdl
        seed1 := 56;
      end if;
      if seed2 < NATURAL'HIGH - 1000 then
        seed2 := seed2 + 1000;
      else
        seed2 := 42;
      end if;
      file_open(seedfile, "seed.dat", WRITE_MODE);
      write(seedfile, seed1);
      write(seedfile, seed2);
      file_close(seedfile);
      return (seed1, seed2);
    end function init_seed;

    variable vseed : seed_pair := init_seed;

    impure function get_seed return seed_pair is
    begin
      return vseed;
    end function get_seed;

    procedure set_seed (sp: seed_pair) is
    begin
      vseed := sp;
    end procedure set_seed;
  end protected body seed;

  shared variable seed : pseed;

  impure function rand return REAL is
    variable seeds : seed_pair;
    variable rnd : REAL;
  begin
    seeds := seed.get_seed;
    uniform (seeds.seed1, seeds.seed2, rnd);
    seed.set_seed(seeds);
    return rnd;
  end function rand;

  function negexp(t : TIME) return TIME is
  begin
    return INTEGER(-log(rand)*(REAL(t / NS))) * NS;
  end function negexp;
end package body random;
```

We can use this random event generator as follows:

```vhdl
library IEEE;
use IEEE.std_logic_1164.all;
use WORK.random.all;
```

```
entity testrnd is end entity testrnd;

architecture testrnd of testrnd is
  signal r : std_logic := '0';
begin
  r <= not r after negexp(10 ns);
end architecture testrnd;
```

12.5.3 VHDL modelling of setup and hold time violations

A structural model of a level-sensitive D latch can be described in VHDL using gate instances or by using a set of concurrent assignments, as shown below. Note that q and qbar are declared as ports with mode **out**, so they cannot be read. Therefore two internal signals, y and z, are used to model the RS latch. If a simulation of this latch is run, using a regular clock and a random event generator for the *D* input, as shown in the testbench fragment, it will be observed that the latch works correctly unless the *D* input changes 2 ns or less before a falling clock edge. If this occurs, the q and qbar outputs oscillate.

Of course, two D latches can be put together to form an edge-triggered flip-flop. The clock input is inverted for the master flip-flop (introducing a delay of, say, 1 ns). Thus, when the clock is low the master flip-flop is conducting. From the previous simulation we would expect therefore that the setup time is 2 ns, less the delay in the clock caused by the inverter, or 1 ns in total. We can verify this by simulation. Again we observe that a change in the D input 1 ns or less before the clock edge may cause the output to oscillate, depending on the state of the flip-flop and whether D is rising or falling. The six-NAND gate edge-triggered D flip-flop behaves similarly. In both cases, the hold time is 0 ns.

```
library IEEE;
use IEEE.std_logic_1164.all;

entity dlatch is
  generic (delay : DELAY_LENGTH := 1 NS);
  port (q, qbar : out std_logic;
        d, c : in std_logic);
end entity dlatch;

architecture concurrent of dlatch is
  signal e, f, g, z, y : std_logic;
begin
  e <= not d after delay;
  f <= d nand c after delay;
  g <= e nand c after delay;
  y <= g nand z after delay;
  z <= f nand y after delay;
  q <= z;
  qbar <= y;
end architecture concurrent;
```

Part of the testbench is shown below.

```
d0 : dlatch port map (q => q, qbar => qbar, d => d,c => c);
d <= not d after negexp(20 NS);
c <= not c after 10 ns;
```

There has to be some doubt as to whether this modelled behaviour is exactly what would be observed in a real circuit. These VHDL models assume that 0 to 1 and 1 to 0 transitions are instantaneous. Of course, in reality, such transitions are finite. Therefore, if a gate had one of its two inputs rising and the other falling simultaneously, it would be reasonable to expect that the output might switch into some state that was neither a logic 1 nor a logic 0 for a period of time. The VHDL standard logic package does not include such a state; 'X' is generally taken to represent a state that could be either 1 or 0.

To show that these results should be treated with caution, we will model the circuit gates in another manner. VITAL (VHDL Initiative Towards ASIC Libraries, 1076.4-2000) allows detailed gate-level timing simulations to be performed by providing a pair of packages for VHDL modelling. A VITAL model of the latch is shown below.

```
library IEEE;
use IEEE.vital_primitives.all;

architecture vtl of dlatch is
  signal e, f, g, z, y : std_logic;
begin
  VitalINV(e, d, (delay, delay));
  VitalNAND2(f, d, c, (delay, delay), (delay, delay));
  VitalNAND2(g, e, c, (delay, delay), (delay, delay));
  VitalNAND2(y, g, z, (delay, delay), (delay, delay));
  VitalNAND2(z, f, y, (delay, delay), (delay, delay));
  q <= z;
  qbar <= y;
end architecture vtl;
```

The references to `VitalINV` and `VitalNAND2` are *concurrent procedure* calls. `VitalINV` and `VitalNAND2` are defined in package `vital_primitives`. A concurrent procedure call is equivalent to a sequential procedure call inside a process with an appropriate wait statement or sensitivity list:

```
process is
begin
  VitalNAND2(a, e, b, (delay, delay), (delay, delay));
  wait on e, b;
end process;
```

The pairs of parameters (`delay`, `delay`) are the delays for a 0 to 1 and a 1 to 0 transition, respectively, for each of the *inputs*. Normally, these two parameters would take different values. Within the VITAL models, rising and falling transitions are considered and appropriate outputs are generated. When this model is simulated, the simulation results are similar to those of the basic NAND gate model, except that after oscillation, because of violation of the fundamental mode restriction, the outputs of the flip-flop go into the 'X' state. Hence two different modelling methods both suggest that the flip-flop behaves in an undesirable way if two inputs change simultaneously, but that the exact nature of that behaviour is uncertain.

12.5.4   Metastability

While the oscillations predicted by both the structural models may occur if the fundamental mode restriction is violated, another condition can occur that a VHDL simulation cannot predict. All flip-flops have two stable states and a third unstable, or *metastable*, state. In this metastable state both flip-flop outputs have an equal value at a voltage level between 0 and 1. A SPICE, or similar, transistor-level operating point analysis is likely to find this metastable condition. This may be likened to balancing a pencil on its point – in theory it is stable, but in practice, noise (vibrations, air movement, etc.) would cause the pencil to topple. The metastable state of a flip-flop is similarly unstable; electrical or thermal noise would cause it to fall into a stable state.

Metastability is most likely to occur when external (asynchronous) signals are inputs to a synchronous system. If metastability is likely to be a problem, then care needs to be taken to minimize its effects. The threat of metastability can never be entirely eliminated, but there is no point in constructing elaborate defences if the chances of its happening are remote. Therefore the critical question is how likely is it to occur? The formula used to calculate the mean time between failures (MTBF) has been found by experiment to be

$$\text{MTBF} = \frac{\exp(T \times t_x)}{f_{clk} \times f_{in} \times T_0}$$

where t_x is the time for which metastability must exist in order that a system failure occurs, f_{clk} is the clock frequency, f_{in} is the frequency of the asynchronous input changes, and T and T_0 are experimentally derived constants for a particular device. If a metastable state occurs at the output of a flip-flop, it will cause a problem if it propagates through combinational logic and affects another flip-flop. Therefore,

$$t_x = t_{clk} - t_{pd} - t_{setup}$$

where t_{clk} is the clock period, t_{pd} is the propagation delay through any combinational logic and t_{setup} is the setup time of the second flip-flop.

Let us put some numbers into this formula. The system is clocked at 10 MHz; therefore t_{clk} is 100 ns. We will examine whether an input flip-flop with a setup time of 10 ns can go into a metastable state, therefore t_{pd} is zero and, hence, t_x is 90 ns. If the asynchronous input changes on average, say, once every 10 clock cycles, f_{in} is 1 MHz. For

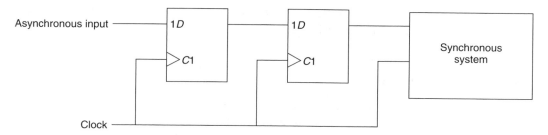

Figure 12.22   Synchronizer design.

a relatively slow D flip-flop (e.g. a 74LS74), T is about 7×10^8 seconds, while T_0 is 0.4 seconds. Therefore

$$\text{MTBF} = \frac{\exp(7 \times 10^8 \times 90 \times 10^{-9})}{10^7 \times 10^6 \times 0.4} = 5.7 \times 10^{14} \text{ seconds}$$

or about 2×10^7 years. Metastability is unlikely to be a problem in such a system! But suppose the clock frequency is doubled to 20 MHz, and hence t_x becomes 40 ns. The asynchronous input changes at the same average rate as before, 1 MHz. Now,

$$\text{MTBF} = \frac{\exp(7 \times 10^8 \times 40 \times 10^{-9})}{2 \times 10^7 \times 10^6 \times 0.4} = 0.18 \text{ second}$$

So we probably will have a problem with metastability in this system.

There are several ways to alleviate the problem. The flip-flop cited above is very slow. A faster flip-flop would have a larger T and a smaller T_0. So using a faster flip-flop will increase the MTBF. Another common solution is to use two flip-flops in series as shown in Figure 12.22. This arrangement does not necessarily reduce the MTBF, but it does reduce the possibility that a metastable state is propagated into the synchronous system.

Although it is fairly unlikely that metastability would be observed in a student laboratory, it is apparent that with increasing clock speeds, and perhaps a move towards a style of design in which there is no global clock, coping with metastability is going to be a challenge for digital designers.

Summary

The design and analysis of asynchronous circuits is harder than for synchronous circuits. Asynchronous circuits may be formally analyzed by breaking feedback loops. The design of an asynchronous circuit starts from a description of all the possible states of the system. A primitive flow table is constructed, which is then minimized. State assignment follows. A poor state assignment can result in race conditions or cycles. From the transition table, next state and output expressions are derived. Hazards can cause erroneous behaviour or oscillations. Essential hazards may result from uneven delays. The design of asynchronous circuits depends on observing the fundamental mode restriction. This is reflected in the specification of setup and hold times for asynchronous blocks used in synchronous design. Failure to observe these restrictions can lead to spurious behaviour and possibly metastability.

Further reading

Although the design of asynchronous (or level-mode, or fundamental mode) sequential circuits is covered in many textbooks, close reading reveals subtle variations in the techniques. Hill and Peterson provide a very good description. Wakerley has a very straightforward description. Unger's 1995 paper has provided perhaps the most rigorous analysis of the problems of metastability. The Amulet project has one of the most significant large asynchronous designs and the website (http://www.cs.man.ac.uk/amulet/index.html) has links to many sources of information about asynchronous design.

Exercises

12.1 What is the difference between a synchronous sequential circuit and an asynchronous sequential circuit? Why is synchronous design preferred?

12.2 What assumption is made in the design of fundamental-mode sequential circuits, and why? How can essential hazards cause the fundamental mode to be violated?

12.3 The excitation equation for a D latch may be written as

$$Q^+ = C.D + Q.\overline{C}$$

Why would a D latch implemented directly from this transition equation be unreliable? How would the D latch be modified to make it reliable?

12.4 Describe, briefly, the steps needed to design an asynchronous sequential circuit.

12.5 Figure 12.23 shows a master–slave edge-triggered D flip-flop. How many feedback loops are there in the circuit, and hence how many state variables?

Derive excitation and output equations and construct a transition table. Identify all races and decide whether the races are critical or non-critical.

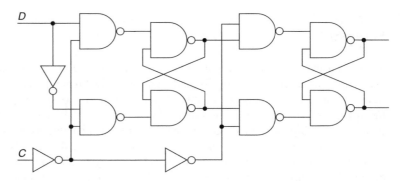

Figure 12.23 Circuit for Exercise 12.5.

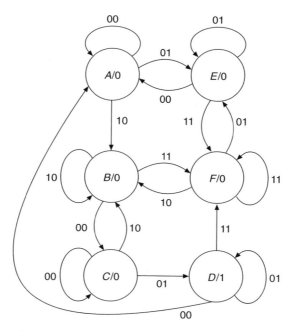

Figure 12.24   State diagram for Exercise 12.6.

Construct a state and output table and show that the circuit behaves as a positive edge-triggered flip-flop.

12.6   Figure 12.24 shows a state diagram of an asynchronous circuit with two inputs, R and P, and a single output, Q. The input values are shown on the arcs; the state names and the output values of the stable states are shown in the circles. Design an asynchronous circuit to implement this function.

12.7   A positive edge-triggered D flip-flop has a preset and clear input, in addition to the clock and D inputs (Figure 12.4). Write down the state equations for the flip-flop *including* the preset and clear inputs. Hence write a transition table.

12.8   Table 12.2 shows the transition table for an asynchronous circuit. Identify all the non-critical races, critical races and cycles (a *cycle* is a repeated series of unstable states that requires an input to change in order for a stable state to be reached).

12.9   Design a D flip-flop that triggers on both the positive and negative edges of the clock pulse.

12.10   An asynchronous sequential circuit has two inputs, two internal states and one output. The excitation and output functions are:

$$Y1^+ = A.B + A.\overline{Y2} + \overline{B}.Y1$$
$$Y2^+ = B + A.\overline{Y1}.\overline{Y2} + \overline{A}.Y1$$
$$Z = B + Y1$$

Table 12.2   Transition table for Exercise 12.8.

Y1 Y2	AB			
	00	01	11	10
00	00	11	10	11
01	11	01	01	10
11	10	11	01	10
10	11	10	01	01
		Y1*Y2*		

(a) Draw the logic diagram of the circuit.

(b) Derive the transition table and output map.

(c) Obtain a flow table for the circuit.

12.11 The **assert** statement shown in Section 12.4 can detect two simultaneous events. Three simultaneous events appear to be the same as one event because of the nature of the **xor** operator. Write an **assert** statement to detect two or three simultaneous events. You may find it easiest to write a function to perform the logical operation. Include the check in a complete model of the asynchronous state machine and verify its operation using a testbench with random events.

Chapter 13

Interfacing with the analogue world

In previous chapters, we have considered the world to be purely digital. Indeed, with the exception of the last chapter, we have further considered only synchronous systems. Of course the real world is asynchronous and, even worse, analogue. All digital systems must at some point interact with the real world. In this chapter, we will consider how analogue inputs are converted to digital signals and how digital signals are converted to analogue outputs. Until recently, the modelling and simulation of digital and analogue circuits and systems would have been performed independently of each other. In 1999, a set of analogue and mixed-signal extensions to VHDL were approved as an IEEE standard (1076.1-1999). The new language is commonly known as VHDL-AMS (Analogue and Mixed-Signal). VHDL-AMS is a complete superset of VHDL. VHDL-AMS simulators that support the entire language are starting to appear, at the time of writing. Having looked at digital to analogue converters (DACs) and analogue to digital converters (ADCs), we will review the basics of VHDL-AMS and see how ADCs and DACs can be modelled in VHDL-AMS. There is not sufficient space to provide a complete tutorial of VHDL-AMS here. Furthermore, it should be remembered that we are considering only simulation models, designed for verifying the interaction of a digital model with the real world. Synthesis of analogue and mixed-signal designs is still a research topic. The final section of the chapter looks at some further mixed-signal circuits and their models in VHDL-AMS.

(13.1) Digital to analogue converters

We will start the discussion of interface circuits with digital to analogue converters because, as we will see, one form of analogue to digital converter requires the use of a DAC. The motivation in this chapter is not to describe every possible type of converter – that would require at least an entire book – but to show one or two examples of the type of circuit that can be employed.

In moving between the analogue and digital worlds, we ideally want to preserve the maximum amount of information. This can be summarized in terms of three aspects: *resolution, accuracy* and *speed. Resolution* defines the smallest change that can be measured. For example, 8 bits can represent 2^8 or 256 voltage levels. If we want to represent a signal that changes between 0 and 5 volts using 8 bits, the resolution is $5/256 = 19.5$ mV. *Accuracy* describes how precisely a signal is represented with respect to some reference. In turn, this depends on factors such as linearity. For example, while 8 bits can represent a 5 V signal with an *average* resolution of 19.5 mV, differences (non-linearities) in the circuit might mean that some changes are really 18.5 mV, while others are 20.5 mV. These differences will add up and affect the overall accuracy. Finally, the *speed* at which data is converted between the two domains affects the design of converters. In the digital world, samples are taken at discrete points in time. The users of converters need to be aware of what happens between these sample points.

The simplest type of DAC is the binary-weighted ladder circuit of Figure 13.1. The bits are added together according to their relative weights. The operational amplifier (opamp) forms a classic (inverting) adder. While this circuit is easy to understand, it is a manufacturing nightmare. The resistors have to be manufactured with very tight tolerances. Any inaccuracy in a resistor value would affect the accuracy. Note that the resistors have to be accurate with respect to the feedback resistor (R), but also with respect to each other.

A variation on the binary-weighted ladder is the R–2R ladder of Figure 13.2. To a significant extent, this overcomes the manufacturing problem as only two values of resistor need to be constructed.

For both these circuits, the speed is limited only by the response of the opamp. In practice, however, we might find that the resistors are more easily implemented as

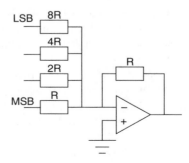

Figure 13.1   Binary-weighted ladder DAC.

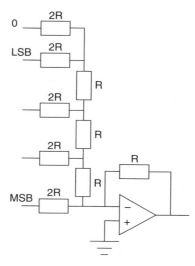

Figure 13.2   Binary-weighted R–2R ladder DAC.

switched capacitors.[1] If this is so, the speed is limited by the clock. Notice also that the output changes in discrete steps.

13.2   Analogue to digital converters

The task of an ADC is to translate a voltage (or current) into a digital code. This is generally harder to achieve than the reverse process. Again, we need to consider resolution, accuracy and speed, but for example suppose we have a signal that changes between 0 V and 5 V, with a maximum frequency of 10 kHz. Eight bits gives a resolution of 19.5 mV, as above. To accurately capture changes in a signal, it needs to be sampled at twice its maximum frequency. Here, therefore, we need to sample at 20 kHz or greater.

Conceptually, the simplest ADC is the flash ADC of Figure 13.3. This consists of nine identical resistors (for eight voltage levels) and eight comparators. As the input voltage, Vin, increases past a level in the resistor change, the corresponding comparator output switches to 1. Therefore, we can use a priority encoder to determine which is the most significant bit, and to encode that value as a binary number. It should be immediately obvious that this circuit is impractical for large numbers of bits. We need 2^n identical, ideal comparators and $2^n + 1$ identical resistors to achieve n bits at the output. It is very difficult to achieve high consistency and hence high accuracy. On the other hand, this type of converter is very fast. In practice, the cost of a flash ADC is usually

---

[1]In CMOS technology, it is generally easier to build accurate capacitors than accurate resistors. It is possible to emulate the behaviour of a resistor by rapidly switching a capacitor between an input and ground.

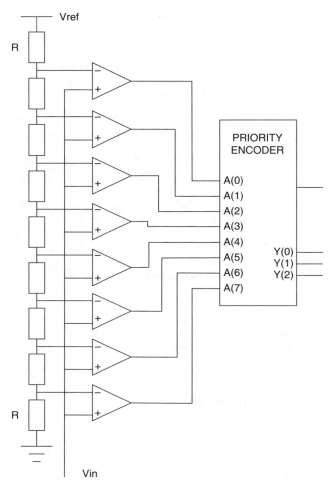

Figure 13.3   Flash ADC.

too high. In return for a smaller design and better accuracy, we pay the price of slower conversion speeds.

Figure 13.4 shows a tracking ADC. This is much easier to implement than the flash ADC. It is essentially a DAC, a comparator and a counter. When the value in the counter is greater than that of the input, Ain, the counter counts down; when the counter's value is less than Ain, the counter counts up. Therefore, the counter attempts to track the input. At first glance, it might appear that a very high clock speed is needed to make this work. Suppose we wish to convert an audio signal with a maximum frequency of 20 kHz. We need to sample at twice this frequency – 40 kHz. In the worst case, the counter needs to count through its entire range, 2^4 or 16 states, between samples. This means that the counter clock must be 16×40 kHz or 640 kHz. On the

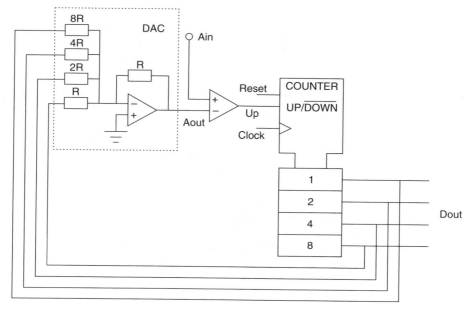

Figure 13.4 Tracking ADC.

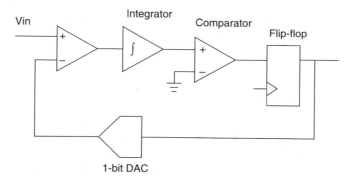

Figure 13.5 Delta-Sigma ADC.

other hand, to achieve CD quality resolution, we would need 16 bits at the output, which implies a clock speed of nearly 3 GHz. This is clearly much less practical.

For high-speed, high-resolution applications an entirely different approach is usually taken. Delta-Sigma ADCs convert from voltage to a serial encoding. Figure 13.5 shows a simple Delta-Sigma ADC. The mark to space ratio of the output is proportional to the ratio of the input voltage to some reference, Vref (as set by the DAC). Let us assume that the DAC output is at Vref. When Vin is less than Vref, the output of the first comparator is negative. This causes the integrator output to ramp downwards. When that output crosses zero (possibly after several clock cycles), the output of the second comparator goes negative. At the next clock edge, a 0 is stored in the flip-flop, causing the DAC to output zero. Now the first comparator causes the integrator to start ramping upwards. Again this might take several clock cycles. In this way, the mark to space ratio of the output is changed. This type of converter is widely used in digital audio applications. The resolution is determined by the clock frequency. As with the tracking ADC, for high resolution, a very high clock speed is needed. However, by using differential coding methods (in other words, by recording changes rather than absolute signal values), the clock speed requirement can be significantly reduced.

In the following section, we will see how some of these circuits can be modelled in VHDL-AMS. It should be borne in mind that these models simply describe the functional behaviour of converters. We have already noted that DACs and ADCs are subject to limitations in terms of accuracy, resolution and speed. Very often it is necessary to model these imperfections and to use the results of such simulations to determine the most suitable designs. As with much else in this chapter, detailed modelling of converter circuits could comprise yet another complete book.

(13.3) VHDL-AMS

VHDL-AMS is a superset of VHDL. Several new keywords and constructs have been added to allow modelling of physical systems. The standard defines the interaction between a standard VHDL simulator and an analogue solver. It is important to realize that VHDL-AMS is not 'analogue VHDL' but a true mixed-signal modelling language. Moreover, VHDL-AMS has been designed to allow modelling of general physical systems, not simply electrical networks.

13.3.1   VHDL-AMS fundamentals

VHDL-AMS introduced some important new concepts. The most important of these can be summed up by the three keywords: **nature**, **terminal** and **quantity**. In 'standard' VHDL, a signal represents a physical wire. When we display the results of a simulation, we can observe the changes of state of that wire over time. Therefore a signal covers two ideas: a physical connection and a time history. In electrical, and other, networks, these two ideas need to be separated. An electrical node represents the point at which two or more components are connected. We cannot, however, talk about the behaviour of that node, unless we specify whether we are referring to its voltage or current or some other aspect.

To distinguish between the physical connection and the behaviour, VHDL-AMS introduces new keywords. The network node is declared as a **terminal**. The voltage or current at that node is declared as a **quantity**. Before giving an example, however, we need to explain how quantities and terminals relate to each other.

A terminal belongs to a particular type of network. For example, an electrical terminal belongs to an electrical network; a magnetic terminal belongs to a magnetic network. Each type of network has behaviour that can be described in terms of quantities. So, for example, the behaviour of an electrical network can be described in terms of voltages and currents, while the behaviour of a magnetic network can be described in terms of magneto-motive force and flux. Each type of network has a pair of quantities. These can be described as through and across quantities. For example, in an electrical network, current flows from one terminal to another *through* network components, while we can also measure the voltage *across* such components. Each type of network has such a pair of through and across quantities. (Note that it is also possible to define an electrical network in which currents are thought of as the across quantities and voltages are thought of as the through quantities. Mathematically, either convention is acceptable. The first convention is more common, however, and we will stick with that. In other kinds of network, the decision about which quantity is across and which is through may be less clear.)

In declaring that a terminal belongs to a particular kind of network, we are effectively defining the through and across quantities for that terminal. Therefore it would not be adequate to declare a terminal to be of a particular VHDL type. Instead, a new construct is used – a **nature**. A nature has three parts: an across type, a through type and a reference node. The reference node is the name of the terminal with respect to which all across quantities are calculated. In electrical networks, this is often known as the ground or earth node.

An electrical nature might be declared as:

```
nature electrical is
    voltage        across
    current        through
    electrical_ref reference;
```

As elsewhere, the keywords are highlighted. Unusually, the keywords **across**, **through** and **reference** follow the identifiers. What are voltage and current? We know that the across and through parts are quantities, so voltage and current must be the types of those quantities. In fact, voltage and current are declared as subtypes of REAL:

```
subtype voltage        is REAL;
subtype current        is REAL;
```

All quantities must be declared as subtypes of REAL. At first glance, this might appear unnecessary. Why not simply say that natures are composed of an across and a through quantity, both of type REAL? There are two reasons for declaring these subtypes. First, by naming the quantities, the meaning becomes clearer. It should be easier to see whether we are doing something silly, such as adding currents and voltages. (VHDL-AMS does not, however, perform dimensional analysis. There would need to be a significant revision of the overall VHDL type system to make that possible.) Second, in VHDL-AMS, we can extend the subtype declaration to include a tolerance:

```
subtype voltage        is REAL tolerance "DEFAULT_VOLTAGE";
```

The string following the tolerance is not part of VHDL-AMS but refers to a parameter passed to the analogue simulator (e.g. the simulator might have a control command such as '.OPTION DEFAULT_VOLTAGE = 1.0E-3'). This defines the accuracy to which quantities of type voltage should be calculated.

In the examples that follow, we will assume that the definitions of the electrical nature are contained in a package, electrical_systems, that has been compiled into the IEEE library.[2]

We can now define one or more terminals with the nature electrical:

terminal node1, node2 : electrical;

Terminals can be declared within architectures (in exactly the same way as signals) or as ports. In port declarations, the keyword **terminal** must be included. Terminal ports have no mode. For example, the entity declaration of a resistor might be:

```
library IEEE;
use IEEE.electrical_systems.all;

entity resistor is
  generic (R: REAL);
  port (terminal node1, node2: electrical);
end entity resistor;
```

At this point, we have created only the physical node. We cannot yet refer to the voltage and current of that node. We now need to declare one or more quantities. There are three types of quantity declaration in VHDL-AMS: free quantities, branch quantities and source quantities. We will not discuss source quantities here – they are used primarily for noise modelling.

A free quantity is simply a declaration of a quantity:

quantity v1 : voltage;

This does not relate v1 to any particular node, and so is not a very useful or likely declaration. More usual would be something like:

quantity q1 : charge;

Free quantities can be given initial values, like variables and signals:

quantity q1 : charge := 0.01;

Branch quantities identify the across and through quantities in some branch between two terminals. So, given the declaration of node1 and node2, above, we can declare:

quantity vr **across** ir **through** node1 **to** node2;

This declares vr to be the voltage between node1 (positive) and node2 (negative), and ir to be the current flowing from node1 to node2. A branch quantity declaration

---

[2]At the time of writing, a number of standard packages were being developed. It is expected that these will be approved as IEEE standard 1076.1.1. We have assumed here that these standard packages have been compiled into the IEEE library.

can be simplified to include only the parts that are needed for a model. The following declares a voltage with respect to ground (`electrical_ref`):

```
quantity v1 across node1;
```

13.3.2   Simultaneous statements

VHDL has three styles of modelling: structural (netlists), concurrent statements and sequential statements (in processes). VHDL-AMS adds two new styles: simultaneous statements and procedurals. We will briefly discuss procedurals later. Simultaneous statements are, in a sense, the analogue equivalents of concurrent statements. Perhaps more accurately, they can be thought of as algebraic equations. Simultaneous statements use the symbol '==' to separate the left and right sides. Note that this is not an assignment. As with algebraic equations, the symbol shows that the two sides should be equal. The left and right sides must each result in a floating point value. Each statement must also contain at least one quantity. (If each side resulted in a constant, the statement would be both unsolvable and irrelevant!)

The simultaneous statements are therefore the simultaneous equations that are solved by the analogue simulator. In accordance with the standard rules of algebra, the number of simultaneous statements should be equal to the number of unknowns, i.e. the free and through quantities. VHDL-AMS has the additional constraint that this equality should be satisfied within each architecture. (This is a slight simplification of the rule, but it is sufficient for the models that we will discuss here.) The across quantities are also unknown, but equations to solve those quantities are created automatically.

In order to illustrate a simultaneous statement, we will give a complete model of a resistor:

```
library IEEE;
use IEEE.electrical_systems.all;

entity resistor is
  generic (R: REAL);
  port (terminal node1, node2: electrical);
end entity resistor;

architecture ohmic of resistor is
  quantity vr across ir through node1 to node2;
begin
  ir == vr/R;
end architecture ohmic;
```

We can model other components in a similar way. For example, a capacitor can be modelled as follows:

```
library IEEE;
use IEEE.electrical_systems.all;
```

```
entity capacitor is
  generic (C: REAL);
  port (terminal node1, node2: electrical);
end entity capacitor;

architecture dvdt of capacitor is
  quantity vc across ic through node1 to node2;
begin
  ic == C*vc'DOT;
end architecture dvdt;
```

The attribute causes a new quantity to be created that is the time derivative of the original quantity. This attribute can be applied only to a named quantity, not to an expression. Similarly, there is an `'INTEG` attribute for calculating the time integral.

Because the simultaneous statement is an algebraic equation and not an assignment, it is not necessary to put the (unknown) through quantity on the left-hand side of the equation. Furthermore, it is possible to write an expression on the left-hand side. If, however, neither the left-hand side nor the right-hand side is a simple quantity, a tolerance clause must be included in the statement to indicate to the solver what accuracy is required in evaluating the expression. All of the following are valid ways to describe a capacitor:

```
ic == C*vc'DOT;
C*vc'DOT == ic;
vc'DOT == ic/C tolerance "DEFAULT_CHARGE";
vc == ic'INTEG/C;
ic'INTEG == C*vc tolerance "DEFAULT_VOLTAGE";
ic'INTEG == C*(node1'REFERENCE - node2'REFERENCE)
            tolerance "DEFAULT_VOLTAGE";
```

In the last example, the attribute `'REFERENCE` is used to obtain the absolute value of each node voltage. If we do this, we do not need to explicitly declare the across quantity vc.

Before leaving these basic models, let us consider a pure voltage source that generates a sine wave. We will need a version of this element to describe a digital to analogue converter.

```
library IEEE;
use IEEE.electrical_systems.all;
use IEEE.math_real.all;

entity vsin is
    generic (vo, va, freq: REAL);
    port (terminal np, nm: electrical);
end entity vsin;

architecture source of vsin is
  quantity vs across iss through np to nm;
begin
    vs == vo + va * sin(2.0*math_pi*freq*NOW);
end architecture source;
```

The `sin` function and `math_pi` are defined in the `math_real` package. In VHDL-AMS the function `NOW` is overloaded to return the current simulation time as a real number (in seconds). Above, we noted that the branch quantity declaration can omit across or through quantities if they are not used. Why, therefore, is the through quantity, `iss`, declared? Recall, also, that for the set of equations to be solvable, the number of through and free quantities (unknowns) must equal the number of simultaneous statements (equations). Therefore, `iss` must be declared, even though it is not explicitly used. This is true of all pure voltage sources.

13.3.3 Mixed-signal modelling

VHDL-AMS is a mixed-signal modelling language. Therefore, we can mix 'analogue' and 'digital' constructs in the same models. Let us consider a simple comparator. We want to convert two analogue voltages into a one-bit digital signal, such that the output is a logic '1' when the first input is greater than the second and '0' otherwise. The entity description can be written as:

```
library IEEE;
use IEEE.std_logic_1164.all;
use IEEE.electrical_systems.all;

entity Comp is
  port (terminal Aplus, Aminus : electrical;
        signal Dout : out std_logic);
end entity Comp;
```

This entity has three ports – two are terminal objects and one is a signal object. The keyword **signal** is optional; it's included here to distinguish the digital port from the analogue ports. This is not a VHDL-AMS extension; we could specify ports to be signals in standard VHDL, but we don't bother, because there is nothing else that they can be.

In the architecture, we need to detect when one voltage becomes greater or less than the other and switch the output accordingly. This could be done with a simple comparison operator, but it is better to use the `'ABOVE` attribute. A threshold, which can be a quantity or an expression, is specified. This creates a new signal and hence causes an event to pass to the digital simulator when the attributed quantity crosses the threshold. Although the attribute is called `'ABOVE`, it causes an event when the threshold is crossed in either direction.

```
architecture Simple of Comp is
  quantity Vplus across Aplus;
  quantity Vminus across Aminus;
begin
  Dout <= '1' when Vplus'ABOVE(Vminus) else
          '0';
end architecture Simple;
```

Notice that there is one concurrent VHDL assignment. There are no through or free quantities, so there are no simultaneous statements.

This example simply converts a signal to one bit. We can use the comparator as part of a flash ADC (see Section 13.2 and Exercise 13.5). Later, we will use the comparator again as part of a tracking ADC. We can, however, also model a flash ADC behaviourally. We simply need to convert a varying (real) quantity into a bit vector. In the example below, the model is parameterized in terms of the analogue voltage range and the number of bits. We also include a clock to sample the waveform – otherwise the model will be evaluated at every analogue time step.

```vhdl
library IEEE;
use IEEE.std_logic_1164.all;
use IEEE.numeric_std.all;
use IEEE.electrical_systems.all;

entity adc is
  generic (Vrange : REAL; N : POSITIVE);
  port (signal Dout : out Std_logic_vector(N-1 downto 0);
        terminal Ain : electrical);
        signal clock : in std_logic);
end entity adc;

architecture Simple of adc is
  quantity Vin across Ain;
  constant Dmax : unsigned(N-1 downto 0) := (others => '1');
begin
  process (clock) is
  begin
    if rising_edge(clock) then
      Dout <=
      std_logic_vector(to_unsigned(INTEGER(Vin/Vrange *
                       REAL(to_integer(Dmax))),N));
    end if;
  end process;
end architecture Simple;
```

In the following example, a digital to analogue converter is modelled as a voltage source and resistance in the analogue world. The voltage source can take one of three values – V1, V0 or Vx for logic 1, logic 0 or unknown, respectively. Similarly, the output resistance can take a low impedance value or a high impedance value. If we assume that metavalues such as 'U' can be mapped onto 'X', this allows us to represent all std_logic values as voltage and resistance pairs. A parameterized entity desciption for a DAC is as follows.

```vhdl
library IEEE;
use IEEE.std_logic_1164.all;
use IEEE.electrical_systems.all;
```

```
entity Dac is
  generic (V1 : REAL := 5.0;
           V0 : REAL := 0.0;
           Vx : REAL := 2.5;
           Zhi : REAL := 1.0e9;
           Zlo : REAL := 1.0);
  port (signal Din : in std_logic;
        terminal Aout : electrical);
end entity Dac;
```

To convert from analogue quantities to digital signals, we write concurrent statements. To convert the other way, we need to write simultaneous statements. There is a catch, however. In discrete simulation ('standard' VHDL), signals change instantaneously. In a continuous simulation, instantaneous step changes cause problems.

Without going into great detail, an analogue or continuous solver approximates a changing quantity by taking discrete time steps. The waveform is therefore approximated by a polynomial expression. The size of these time steps is varied to minimize the error in the polynomial. A large step change makes it impossible to construct a polynomial expression across that change, so the error is considered large and the time step is reduced in an attempt to minimize the error. No matter how small the time step is made, the error will remain large and the simulation fails.

One way to avoid instantaneous changes is to force a transition to occur in a finite time. This can be done with the 'RAMP attribute. The values of the voltage and resistance are held as signals within the DAC model. When the input signal changes, these signals are updated. An expression for the output voltage in terms of these signals can then be written as a simultaneous statement. Note that changes in the signals are slowed by 1 ns using the 'RAMP attribute in the simultaneous statement.

```
architecture ramped of Dac is
  quantity Vout across Iout through Aout;
  signal Zth : REAL := Zlo;
  signal Vth : REAL := Vx;
begin
  Zth <= Zhi when Din = 'Z' else
         Zlo;
  Vth <= V0 when Din = '0' else
         V1 when Din = '1' else
         Vx;
  Vout == Vth'RAMP(1.0e-9) - Iout*Zth'RAMP(1.0e-9);
end architecture ramped;
```

The obvious disadvantage of this approach is that the time to change between values has to be specified; 1 ns might easily be far too large or far too small compared with other changes in the system. It would be better to let the solver decide for itself what would constitute a suitable change. For this to happen, the solver needs to be told that there could be a problem, and this is the responsibility of the model writer. VHDL-AMS includes a mechanism for indicating a discontinuity – the **break** statement.

```
architecture breakon of Dac is
  quantity Vout across Iout through Aout;
begin
  break on Din;
  case Din use
    when '0' => Vout == V0 - Iout*Zlo;
    when '1' => Vout == V1 - Iout*Zlo;
    when 'Z' => Vout == Vx - Iout*Zhi;
    when others => Vout == Vx - Iout*Zlo;
  end case;
end architecture breakon;
```

When Din changes, the analogue solver stops and restarts, therefore avoiding the error detection mechanism. In this example, we also introduce the *simultaneous case* statement. Unlike the sequential case statement, this exists outside a sequential block. Each branch of the **case** statement consists of one simultaneous statement. At any time, only one of these statements is therefore valid and thus the solvability conditions are met. The syntax of the **case** statement is very similar to that of the sequential case statement, except for the inclusion of **use** instead of **is**. There is also a simultaneous **if** statement, which we will introduce later.[3]

If we wish to convert several bits to an analogue equivalent, we could use a one-bit DAC for each input bit and add the outputs together, with appropriate weighting. If we are not concerned with converting 'X' and 'Z' bits, it is easier to simply convert the bits to a real number as follows. Notice that the output is scaled to a generic, Vref.

```
library IEEE;
use IEEE.std_logic_1164.all;
use IEEE.numeric_std.all;
use IEEE.electrical_systems.all;

entity NbitDac is
  generic (Vref : REAL; N : POSITIVE);
  port (signal Din : in std_logic_vector(N-1 downto 0);
        terminal Aout : electrical);
end entity NbitDac;

architecture Nbit of NbitDac is
  quantity Vout across Iout through Aout;
  constant Dmax : unsigned(N-1 downto 0) := (others => '1');
begin
  break on Din;
  Vout == (Vref*REAL(to_integer(unsigned(Din))))/
          REAL(to_integer(Dmax));
end architecture Nbit;
```

---

[3]There is also an inconsistency in the syntax here; the last line of a simultaneous case statement is **end case**. The last line of a simultaneous **if** statement is **end use**. Don't try to memorize this – that is what compilers are for!

We now have the necessary parts to build the tracking ADC from Section 13.2. We also need the counter from Exercise 6.6. This has been written as a self-contained testbench. It would be equally valid to include the four component parts in a separate entity. Notice that we have created a netlist in exactly the same way as digital netlist, the only difference being that the analogue nodes needed for connecting components are declared as **terminal**s.

```
library IEEE;
use IEEE.std_logic_1164.all;
use IEEE.numeric_std.all;
use IEEE.electrical_systems.all;

entity Tracking is
end entity Tracking;

architecture Structural of Tracking is
  signal Clock, Reset, Up : std_logic := '0';
  signal Dout : std_logic_vector(3 downto 0);
  terminal Ain, Aout : electrical;
begin
  Clock <= not Clock after 10 NS;
  Reset <= '1', '0' after 2 NS;
  C1 : entity WORK.Counter generic map (4)
                           port map (Clock, Reset, Up,
                                     Dout);
  D1 : entity WORK.NBitDac generic map (5.0, 4)
                           port map (Dout, Aout);
  O1 : entity WORK.Comp port map (Ain, Aout, Up);
  V1 : entity WORK.Vsin generic map (2.5, 2.5, 1.0E4)
                        port map (Ain, electrical_ref);
end architecture Structural;
```

13.4 Phased-locked loops

Although ADCs and DACs are the main interfaces between the analogue and digital worlds, another class of circuits also sits at this boundary. One of the major uses for phase-locked loops (PLLs) is for generating clocks. PLLs can be used to recover the clock from a stream of data. A PLL can also be used to 'clean up' a clock that has an irregular period and to multiply a clock signal to create a higher frequency signal. All of these tasks are difficult to achieve with conventional digital circuit techniques. PLLs can be built as purely analogue circuits, as purely digital circuits or using a mixture of methods. As with ADCs and DACs, there is not enough space in a book like this to give any more than a brief introduction to PLLs. The purpose here is to show a simple example and to illustrate one way of modelling that example in VHDL-AMS. As with ADCs and DACs, the real art of modelling PLLs is to capture non-linearities and other imperfections to determine whether a particular design will work in a particular context.

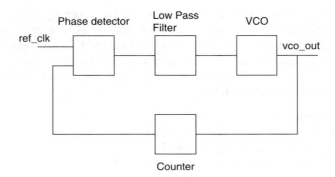

Figure 13.6   PLL structure.

Figure 13.6 shows the basic structure of a PLL. The phase detector determines the difference between the input (ref_clk) and the stabilized output (vco_out). The phase detector could be an analogue four-quadrant multiplier or a digital XOR gate or a sequential digital circuit. The output from the phase detector is a sequence of pulses. The low pass filter averages these pulses in time. This filter is important to the working of the PLL. If its time constant is too small, the PLL will not settle into a regular 'locked' pattern. If the time constant is too great, the PLL may not lock at all. The voltage controlled oscillator (VCO) converts the output of the filter into a oscillation whose frequency is determined by the filter output voltage. The VCO is likely to be the hardest part of the design. It can oscillate only within a relatively narrow band of frequencies. Finally, the counter is optional. By dividing the VCO output, the phase detector compares with this reduced frequency output. In other words the VCO output must be a multiple of the input frequency.

There are many books about PLL design, but perhaps the best way to understand their operation is by playing with the circuit parameters in a simulation. Therefore, we will simply present one, ideal, model of a PLL.

We start with the largest model – the phase detector. We will use the example from Chapter 12. The two outputs, qa and qb, correspond to two control signals, up and down, respectively. These need to be converted to analogue voltages and filtered. We will use two instances of the one-bit DAC from the previous section. The low pass filter can be modelled using the Laplace Transform attribute in VHDL-AMS. This attribute takes two parameters, each of which is a vector of real numbers. The first vector contains the coefficients of the numerator and the second contains the coefficients of the denominator. Here we want to create a parameterized low pass filter, which in the s-domain has the transfer function $\frac{1}{1 + sT}$.

Therefore the numerator has the value 1.0, and the denominator has the values 1.0 and T. Hence, this is the VHDL-AMS model. Although this is a frequency domain model, it can be interpreted in the time domain. Similarly, time domain models (such as v'DOT) can be interpreted in the frequency domain.

```
library IEEE;
use IEEE.electrical_systems.all;
```

```
entity lpf is
  generic (T : REAL := 1.0e-6);
  port (terminal Ao, Ai : electrical);
end entity lpf;

architecture ltf of lpf is
  constant Num : REAL_VECTOR := (0 =>1.0);
  constant Den : REAL_VECTOR := (1.0, T);
  quantity Vo across Io through Ao;
  quantity Vi across Ai;
begin
  Vo == Vi'LTF(Num, Den);
end architecture ltf;
```

The voltage controlled oscillator is mixed-signal, but can be written using a VHDL process.

```
library IEEE;
use IEEE.Std_logic_1164.all;
use IEEE.electrical_systems.all;

entity vco is
  generic (gain : REAL := 5.0e5;
           fnom : REAL := 2.5e5;
           vc : REAL := 2.5);
  port (terminal Ina, Inb : electrical;
        vout : out std_logic);
end entity vco;

architecture a2d of vco is
begin
  process is
    variable frequency : REAL;
    variable period : REAL;
  begin
    frequency := fnom +
      (Ina'REFERENCE - Inb'REFERENCE - vc)*gain;
    if frequency > 0.0 then
      period := 1.0/frequency;
    else
      period := 1.0/fnom;
    end if;
    wait for period/2.0;
    vout <= '1';
    wait for period/2.0;
    vout <= '0';
  end process;
end architecture a2d;
```

Note that there are VHDL-AMS extensions, even within the process. The values of the input quantities are found using the 'REFERENCE attribute and the **wait** statements take real numbers, not time units.

The counter is purely digital, although we will use it in an asynchronous way.

```vhdl
library IEEE;
use IEEE.std_logic_1164.all;

entity counter is
  generic(n : NATURAL := 4);
  port(clk : in std_logic;
       count : out std_logic);
end entity counter;

architecture rtl of counter is
begin
p0: process (clk) is
      variable cnt : NATURAL;
    begin
      if clk = '1' then
        cnt := cnt + 1;
        if cnt = n then
          cnt := 0;
          count <= '1';
        else
          count <= '0';
        end if;
      end if;
    end process p0;
end architecture rtl;
```

Finally, we can put all the parts together and include a suitable stimulus.

```vhdl
library IEEE;
use IEEE.std_logic_1164.all;
use IEEE.electrical_systems.all;

entity pll is
end entity pll;

architecture structural of pll is
  terminal up, down, up_a, down_a : electrical ;
  signal up_d, down_d, VCO_out, VCO_div : std_logic ;
  signal ref_clk : std_logic := '0' ;
begin
  P0: entity WORK.phase_detector port map (inA => ref_clk,
          inB => VCO_div, qA => up_d, qB => down_d);
```

```
D0: entity WORK.dac port map (Din => up_d, Aout => up_a);
D1: entity WORK.dac port map (Din => down_d, Aout =>
                              down_a);
L0: entity WORK.lpf generic map (50.0e-6)
        port map (Ai => up_a, Ao => up);
L1: entity WORK.lpf generic map (50.0e-6)
        port map (Ai => down_a, Ao => down);
V0: entity WORK.vco generic map (gain => 1.0e5,
        fnom => 8.0e5, vc => 2.5)
        port map (Ina => down, Inb => up, vout =>
                  VCO_out);
C0: entity WORK.counter generic map (5)
        port map (clk => VCO_out, count => VCO_div);
ref_clk <= not ref_clk after 5 US;
end architecture structural;
```

Simulation of this PLL model shows that the output frequency varies between about 450 kHz and 600 kHz, before settling at 500 kHz after about 260 μs. The clock has a frequency of 100 kHz and the counter counts to 5, so the PLL behaves exactly as we would expect.

13.5 VHDL-AMS simulators

It could be argued that the mixed-signal models of ADCs, DACs and PLLs could be modelled entirely in standard VHDL. Indeed, there is a very limited amount of behaviour that requires an analogue solver in these models. The real power of VHDL-AMS is that it allows digital VHDL models to be simulated at the same time as analogue circuits that would traditionally have been simulated with SPICE. For several reasons, it is appropriate to bring the discussion to a close.

First, at the time of writing, there are very few VHDL-AMS simulators available. None of these yet cover the entire language. For example, there is a whole style of modelling that is still unsupported by most of these tools. Just as it is possible to write sequential code in 'digital' VHDL, it is possible to write sequential code in VHDL-AMS. For example, the following is a model of an opamp that uses the **procedural** construct.

```
library IEEE;
use IEEE.electrical_systems.all;

entity Opamp is
  generic (Rin        : REAL := 4.0E5;
           Vinoffset  : REAL := 40.0E-6;
           Rout       : REAL := 640.0;
           Gain       : REAL := -50.0;
           Voutoffset : REAL := 0.0;
           Iddtf      : REAL := 70.0E-6);
```

```
  port(terminal Inn, Outt, Psu : electrical);
end entity Opamp;

architecture ClosedLoop of Opamp is
  quantity Vout across Iout through Outt;
  quantity Vin across Iin through Inn;
  quantity Vdd across Idd through Psu;
begin
  procedural is
    variable Vin1, Vo : Voltage;
  begin
    Vin1 := Vin - Vinoffset;
    Vo := Vin1 * Gain + Voutoffset;
    Iin := Vin1 / Rin;
    Vout := Vo - (-Iout * Rout);
    Idd := -((Vin*Iddtf) + Iout);
  end procedural;
end architecture ClosedLoop;
```

This model will compile in some simulators, but will not simulate. You may also find that parts of the 1993 VHDL standard are not implemented in VHDL-AMS simulators.

Second, the VHDL-AMS standard does not specify an algorithm for the analogue solver. At present, most VHDL-AMS simulators are based on a SPICE-like simulator. There is a particular problem in calculating initial conditions as assumptions may be built into the solver. For example, in one simulator, the n-bit DAC needed the following modification:

```
if domain = quiescent_domain use
  Vout == 0.0;
else
  Vout == (Vref*REAL(to_integer(unsigned(Din))))/
          REAL(to_integer(Dmax));
end use;
```

DOMAIN is a built-in signal that can take the values QUIESCENT_DOMAIN, TIME_DOMAIN or FREQUENCY_DOMAIN. The signal is automatically updated as each type of analysis is performed. Similar 'fixes' may be needed in other models. It may also be necessary to change the **library** and **use** clauses to make these models work with particular simulators.

Another compatibility problem exists with SPICE models. It is very likely that the analogue part of a design will be described in terms of SPICE models, either at component level or in terms of SPICE macromodels. Although SPICE is widely used, it is not standardized. Therefore each simulator accepts a slightly different set of components and commands. To use SPICE models in a VHDL-AMS description, either there needs to be an interface between VHDL-AMS and SPICE (which is not defined and therefore unique to each simulator) or the SPICE models need to be converted to VHDL-AMS (which in turn raises a host of compatibility issues). Nevertheless, both these approaches can be used in existing VHDL-AMS simulators. Therefore, for the immediate future, it is likely that the use of VHDL-AMS will be limited to providing interfaces between VHDL and SPICE.

Summary

At some point, digital circuits have to interface with the real, analogue world. Modelling this interface and the interaction with analogue components has always been difficult. VHDL-AMS extends VHDL to allow analogue and mixed-signal modelling. Typical converters include ladder DACs, flash ADCs, Delta-Sigma ADCs and PLLs. All of these components can be modelled and simulated in VHDL-AMS. There is, as yet, no way to automatically synthesize such elements from a behavioural description. VHDL-AMS simulators are still relatively new and generally do not support the entire language. They do, however, provide means for interfacing between SPICE models and VHDL-AMS, allowing modelling of complete systems.

Further reading

For an explanation of analogue simulation algorithms, see Litovski and Zwolinski. Horowitz and Hill is an excellent guide to practical circuit design and includes descriptions of ADCs, DACs and PLLs. For a full description of VHDL-AMS, the language reference manual is, of course, invaluable. Manufacturers' manuals need to be read with the LRM to understand any limitations.

Exercises

13.1 An inductor is described by the equation $v_L = L \cdot \dfrac{di_L}{dt}$. Write a VHDL-AMS model of an inductor, using the `'DOT` attribute.

13.2 Write an inductor model that uses the `'INTEG` attribute.

13.3 Write a parameterizable model of a voltage source that generates a ramp. The parameters should be initial voltage, final voltage, delay before the ramp, and rise (or fall) time.

13.4 Write a model of voltage source that generates a pulse. What parameters need to be specified? How is it made to repeat?

13.5 Write a VHDL-AMS model of the flash ADC shown in Figure 13.1.

VHDL standards

A.1 VHDL and related standards

IEEE Std 1076-2002

IEEE Standard VHDL Language Reference Manual

This is the formal definition of VHDL. The original version of the standard was produced in 1987. A revised version of the standard was produced in 1993 and this version is implemented in most tools. The differences between the 2002 and 1993 versions are slight, with the exception of shared variables. Some tools still conform only to the 1987 version. Sections A.2 and A.3 summarize the differences between the 1987, the 1993 and the 2002 standards. Appendix C lists alternative forms of the packages used in this book that use shared variables.

IEEE Std 1076.1-1999

Analog and mixed signal extensions to IEEE Standard VHDL

VHDL was designed for describing digital systems. IEEE Std 1076.1, known colloquially as VHDL-AMS, defines a set of extensions to allow analogue and mixed-signal modelling and simulation.

IEEE Std 1076.2-1996

IEEE standard VHDL mathematical packages

Most programming languages have built-in functions for performing floating-point mathematical operations. This standard defines real and complex functions for VHDL.

IEEE Std 1076.3-1997

IEEE standard VHDL synthesis packages

These two packages (for types `bit` and `std_logic`, respectively) define the `signed` and `unsigned` types and arithmetic functions, for use with synthesis tools. These packages replace a number of vendor-specific packages that achieved the same results.

IEEE Std 1076.4-2000

IEEE standard VITAL application-specific integrated circuit (ASIC) modeling specification

VITAL (VHDL Initiative Towards ASIC Libraries) is a set of low-level primitives for accurate timing simulations of gate-level models.

IEEE Std 1076.6-1999

IEEE standard for VHDL register transfer level synthesis

The RTL synthesis standard defines the subset of VHDL appropriate to RTL synthesis. Certain constructs, for example delays and floating-point operators, are not synthesizable. This standard defines what a synthesis tool should be able to implement. The first version of this standard recognizes only VHDL conforming to the 1076-1987 standard. VHDL'93 constructs may be ignored.

IEEE Std 1029.1-1998

IEEE standard for waveform and vector exchange (WAVES)

WAVES (Waveform and Vector Exchange to Support Design and Test Verification) is a set of VHDL methods to assist in verifying and testing hardware. The motivation for this standard is to allow test vector files to be shared between VHDL simulators and hardware testers.

IEEE Std 1164-1993

IEEE standard multivalue logic system for VHDL model interoperability

This defines the `std_ulogic` and `std_logic` types, together with the respective vector types and their Boolean operators.

A.2 Differences between 1076-2002 and 1076-1993

The 2002 standard is essentially the same as the 1993 standard, but certain ambiguities have been removed. Most of the examples in this book conform to both standards. There are a small number of differences as follows.

The restriction on the use of ports with mode **buffer** has been eased. In the 1993 standard, a port signal with this mode can be written to and read within an architecture, but can be connected only to an internal signal or a port also with mode **buffer** within an instantiating architecture. This restricts how such a model can be used. The reason for this restriction was that a signal with mode **buffer** may have only a single driver. In the 2002 standard, ports with mode **buffer** can be connected to ports with mode **out** or **inout** in the instantiating architecture. Similarly, ports with mode **out** or **inout** can be connected to ports with mode **buffer** in the instantiating architecture. This makes it possible to use the **buffer** mode for modelling devices such as latches, without requiring internal signals.

Thus, the following is legal in the 2002 standard, but not in earlier versions.

```
entity nand2 is
  port (z : out BIT; a, b : in BIT);
end entity nand2;
architecture dataflow of nand2 is
begin
  z <= a nand b;
end architecture dataflow;

entity latch is
  port (q, qb : buffer BIT; r, s : in BIT);
end entity latch;
architecture gate of latch is
begin
g0: entity WORK.nand2 port map (q, qb, r);
g1: entity WORK.nand2 port map (qb, q, s);
end architecture gate;

entity dlatch is
  port (q, qb : out BIT; d, c : in BIT);
end entity dlatch;
architecture struct of dlatch is
  signal notd, r, s : BIT;
begin
  notd <= not d;
n0: entity WORK.nand2 port map (r, d, c);
n1: entity WORK.nand2 port map (s, notd, c);
l0: entity WORK.latch port map (q, qb, r, s);
end architecture struct;
```

The function NOW, which returns the current simulation time, is now defined as a pure function, not an impure function. In the 1987 standard, there were no impure functions! This has implications for the definition of functions that themselves use NOW.

Shared variables are discussed in Appendix C.

The following features are marked as likely to be removed in the next revision (which is why they have been ignored in this book). It is suggested that these features should not be used for new VHDL code.

1. The **linkage** mode for ports.

2. Replacement characters. '**!**' can replace '**|**' (**when** statements); '**%**' can replace '**"**' (strings); and '**:**' can replace '**#**' in delimiting numbers.

(A.3) Differences between 1076-1993 and 1076-1987

The 1993 VHDL standard was largely an extension of the 1987 standard. The examples in this book conform to the 1993 standard and will not work unmodified with some older tools. With the exceptions of file handling and of impure functions not explicitly declared as such, VHDL written according to the 1987 standard will compile with tools conforming to the 1993 standard.

The following new reserved words were added to the 1993 standard: **group**, **impure**, **inertial**, **literal**, **postponed**, **pure**, **reject**, **rol**, **ror**, **shared**, **sla**, **sll**, **sra**, **srl**, **unaffected** and **xnor**.

Most of the other changes from the 1987 to 1993 standards are rationalizations of various structures. Listed below are some of the changes that would have to be made to the examples given, to make them acceptable to a 1987 compliant tool:

```
end entity XYZ;
end architecture XYZ;
end configuration XYZ;
end procedure XYZ;
end function XYZ;
end package XYZ;
end package body XYZ;
component XYZ is
process XYZ is
```

The form given here is more verbose, but easier to read. Note also that labels can be prepended to most VHDL constructs in the 1993 standard. The 1987 places much tighter restrictions on which constructs may be labelled.

Here is an example of a file declaration, using the 1987 standard:

```
use std.textio.all;
file DataFile : text is in "data.dat";
```

This is how it would be done using the 1993 standard:

```
use std.textio.all;
file DataFile : text open read_mode is "data.dat";
```

In addition, the procedures `file_open` (two forms) and `file_close` are implicitly defined when a new file type is declared.

The 1987 standard had two attributes that worked with blocks or architectures: `'STRUCTURE` and `'BEHAVIOR`. These have been deleted. The following attributes were added in the 1993 standard: `t'ASCENDING`, `t'IMAGE(x)`, `t'VALUE(x)`, `a'ASCENDING`, `s'DRIVING`, `s'DRIVING_VALUE`, `e'SIMPLE_NAME`, `e'INSTANCE_NAME` and `e'PATH_NAME`.

(A.4) VHDL 200x

The next revision of VHDL is likely to be a lot more radical than the 2002 revision. Various groups around the world are considering matters such as object-oriented extensions to VHDL and including the `std_logic` package as part of the main standard. At this stage it is not possible to predict how and when these changes might finally appear. In any case, VHDL like all IEEE standards is subject to a balloting process during which significant syntactic and semantic changes could emerge.

For example, one model of object-oriented VHDL (SUAVE) would allow types to be passed to entities as, for example:

```
entity counter is
  generic (type count_type is (<>));
  port (clk : in BIT;
        data : out count_type);
end entity counter;
```

This entity declaration would allow the same counter model to be used, for example, for both `natural` and `unsigned` types. A number of other extensions are proposed in SUAVE.

Another approach (Objective VHDL) allows one declaration to be derived from another. For example, a flip-flop with an enable signal could be derived from a basic flip-flop as follows:

```
entity DFF is
  port(D, Clk : in bit;
       Q : buffer bit);
end entity DFF;

entity EDFF is new DFF with
  port(EN : in Bit);
end entity EDFF;
```

Again this is only one enhancement among many.

Verilog

Verilog is often seen as an alternative to VHDL. The Verilog language was developed in the early 1980s by Gateway Design Automation, which was later taken over by Cadence. Verilog was then put into the public domain and became an IEEE standard (1364-1995 and 1364-2001). In many respects, Verilog resembles the C programming language, while VHDL is closer to Ada. Verilog is weakly typed: assignments that would be illegal in VHDL are permissible in Verilog. Verilog is often said to be simpler and closer to hardware than VHDL. Indeed, Verilog can be used to model logic circuits at the transistor or switch level, which is difficult in VHDL. Similarly, fault simulators have been developed to use Verilog, while such tools are almost non-existent for VHDL. On the other hand, VHDL has high-level constructs and abstract data types that Verilog does not have, making VHDL much more suited to behavioural modelling. Many simulation and RTL synthesis tools accept both VHDL and Verilog. Therefore it is possible to use VHDL for high-level design and Verilog for low-level post-synthesis timing and fault simulation.

This appendix is a brief introduction to Verilog. The purpose is to highlight the differences (and similarities) between VHDL and Verilog, not to provide a detailed tutorial.

A Verilog model of a two-input NAND gate is shown below.

```
// Comment
module NAND (in1, in2, out);
   input in1, in2;
   output out;
   assign out = ~(in1 & in2);
endmodule // Note no semicolon
```

Verilog is case-sensitive (like C, but unlike VHDL). Identifiers must start with a letter or underscore, and can include letters, digits, _ and $. Note that Verilog does not have separate interface and implementation sections – everything is in a **module**. Signals do not have types; a signal can take the value 0, 1, Z or X. **assign** introduces a continuous assignment statement, analogous to a concurrent assignment in VHDL.

The bitwise logical operators are ~ (NOT), & (AND), | (OR) and ^ (XOR). To instantiate a module, it simply has to be invoked:

```verilog
module simple (a, b, c, d);
  input a, b, c;
  output d;
  wire p, q;
  NAND g1 (a, b, p);
  NAND g2 (a, c, q);
  NAND g3 (p, q, d);
endmodule
```

g1 to g3 are instance names. The internal signals, p and q, are declared with **wire**, although this declaration can be omitted (the compiler would work out that p and q are wires!). In order for the compilation to succeed, the NAND module must be compiled first.

As with VHDL, the sequential programming constructs are much more powerful than the concurrent constructs. For example, a 4 to 1 multiplexer could be modelled by:

```verilog
module MUX (Sel, A, B, C, D, Y);
  input [1:0] Sel; // 2 bit vector
  input A, B, C, D;
  output Y;
  reg Y;              // needed for procedural assignment
  always @(Sel or A or B or C or D)
    case (Sel)
      2'b00 : Y = A;
      2'b01 : Y = B;
      2'b10 : Y = C;
      2'b11 : Y = D;
      default : Y = A;
    endcase
endmodule
```

The procedural (sequential) section is introduced by the **always** reserved word. This is analogous to a VHDL **process**. The **always** section is executed when any of the inputs in the expression following the @ changes. The sequential section has only one statement: the **case** statement. If it had more than one statement, we would need to bracket the statements with **begin** and **end**. The constants in the case statement are of the form 2'b00, where the first digit shows the number of bits and the b shows that the digits are binary. The characters o (octal), h (hexadecimal) and d (decimal) may also be used.

A level-sensitive latch can be modelled using an incomplete **if** statement, as in VHDL:

```verilog
module LATCH (En, D, Q);
  input En, D;
  output Q;
  reg Q;
  always @(En or D)
```

```
    if (En)
        Q <= D;
endmodule
```

There are two sorts of procedural assignment: *blocking* and *non-blocking*. Two non-blocking assignments are shown in the following code fragment:

```
reg M;
always @(En or D)
  if (En)
    begin
    M <= D;
    Q <= M;
    end
```

Like a VHDL signal assignment (which it syntactically resembles), all non-blocking assignments are completed at the end of the current time period. (Note that Verilog does not have the concept of delta delays.) Therefore the example shown would synthesize to two latches. On the other hand, the following code fragment has blocking assignments and would synthesize to a single latch.

```
reg M;
always @(En or D)
  if (En)
    begin
    M = D;
    Q = M;
    end
```

A blocking assignment must be completed before control passes to the next statement. This is similar to a variable assignment in VHDL (but there is no distinction between signals and variables in Verilog). In general, to write portable code for synthesizable hardware, blocking assignments should be used for combinational logic and non-blocking assignments used for sequential hardware. The two types of assignment should not be mixed in the same procedural block.

An edge-triggered flip-flop is modelled by detecting the edge:

```
module DFF (Clk, D, Q);
  input Clk, D;
  output Q;
  reg Q;
  always @ (posedge Clk)
      Q <= D;
endmodule
```

A negative edge would be detected using **negedge**. Similarly, a flip-flop with an asynchronous reset would detect an edge on the reset:

```
module DFF (Clk, Reset, D, Q);
  input Clk, Reset, D;
```

```verilog
  output Q;
  reg Q;
  always @(posedge Clk or negedge Reset)
    if (!Reset)
      Q <= 0;
    else
      Q <= D;
endmodule
```

Verilog does not have enumerated types. Therefore state machines require the state assignment to be explicitly stated. The integer values would be mapped directly onto binary values in synthesis. The following example shows a Verilog version of the vending machine example from Chapter 5, followed by a suitable testbench.

```verilog
module
vending(clock,reset,twenty,ten,ready,dispense,ret,coin);
  input clock,reset,twenty,ten;
  output ready,dispense,ret,coin;
  reg ready,dispense,ret,coin;
  parameter A=0, B=1, C=2, D=3, F=4, I=5;
  reg [0:2] present_state, next_state;
always @(posedge clock or posedge reset)
  if (reset)
    present_state <= A;
  else
    present_state <= next_state;
always @(twenty or ten or present_state)
  begin
    ready = 0;
    dispense = 0;
    ret = 0;
    coin = 0;
    case (present_state)
      A : begin
            ready = 1;
            if (twenty)
              next_state = D;
            else if (ten)
              next_state = C;
            else
              next_state = A;
          end
      B : begin
            dispense = 1;
            next_state = A;
          end
```

```verilog
      C : begin
            coin = 1;
            if (twenty)
              next_state = F;
            else if (ten)
              next_state = D;
            else
              next_state = C;
            end
      D : begin
            coin = 1;
            if (twenty)
              next_state = B;
            else if (ten)
              next_state = F;
            else
              next_state = D;
            end
      F : begin
            coin = 1;
            if (twenty)
              next_state = I;
            else if (ten)
              next_state = B;
            else
              next_state = F;
            end
      I : begin
            ret = 1;
            next_state = A;
            end
      default : next_state = A;
    endcase
  end //always
endmodule

`timescale 1ns / 100ps

module testbench;
  reg clock, reset, twenty, ten;
  wire ready, dispense, ret, coin;
  vending vm
  (clock,reset,twenty,ten,ready,dispense,ret,coin);
  always
    begin
    #10 clock = 1'b0;
```

```
      #10 clock = 1'b1;
      end
   initial
      begin
      reset = 1;
      twenty = 0;
      ten = 0;
      #1  reset = 0;
      #64 twenty = 1;
      #80 twenty = 0;
      #20 ten = 1;
      #20 ten = 0;
      #20 twenty = 1;
      #20 twenty = 0;
      end
endmodule
```

The reserved word **initial** introduces a sequential block that is executed once. Hence it is suitable for generating inputs in testbenches. The notation #10 indicates a delay of 10 time units. The directive 'timescale 1ns/100ps states that the time units are nanoseconds, with a resolution of 100 ps.

In summary, there are a lot of features of the Verilog language that we have not discussed here. Nevertheless, the main advantages of Verilog are that it is relatively concise and that low-level hardware modelling is easier than in VHDL. Compared with VHDL its disadvantages are the lack of abstract data types, making behavioural modelling difficult; weak type checking, which makes it easy to write poor code; and the absence of the delta delay model, meaning that simulation results may vary between simulators. Nevertheless, it is becoming increasingly important that hardware designers are able to use, or at least understand, both VHDL and Verilog.

Shared variable packages

The **shared variable** construct was introduced in 1993. This construct caused some disquiet when it was introduced, because it can cause simulations to be non-deterministic. For example, the order in which the fault list is built in the fault simulator code of Section 10.5.3 may vary between simulators. When a VHDL simulation starts, all processes are executed once until the first **wait** statement (which may be implicit), when they suspend. The order of execution of the processes is not defined. Normally this does not matter. A **shared variable** can, however, be read from or written to in different processes. Unlike the situation with a signal, a process would not suspend while it was waiting for a shared variable. Thus the value of x in the example below could be '0' or '1' at the beginning of the simulation, depending on which process was executed first (which may in turn depend on which process was compiled first!).

```
shared variable sv : BIT := '0';
signal x : BIT;
a : process is
    begin
        sv := '1';
        wait for 10 NS;
    end process a;
b : process is
    begin
        x <= sv;
        wait for 5 NS;
    end process b;
```

In late 1999, a revision to the VHDL standard (1076A) was approved. This significantly changed the way in which shared variables can be used. The change was incorporated into the 2002 revision to VHDL.

In the new standard, to give the user control, the types of shared variables are declared as **protected** (a new reserved word). Within a protected type declaration, variables, functions and procedures may be declared. Protected types and the publicly visible parts of declared types may not, however, be access or file types. The most important point to note is that functions and procedures have to be used to change the value of the field within a protected type.

The fault simulator of Section 10.5.3 uses shared variables and needs to be written in two ways for the two versions of the standard. This is the full VHDL 1076-2002 compliant package definition.

```vhdl
package fault_inject is
  type fault_list is protected
    impure function new_fault(name : STRING)
      return NATURAL;
    procedure first_fault;
    impure function end_fault_list return BOOLEAN;
    procedure inc_fault_list;
    impure function simulating(fault_no : NATURAL) return
      BOOLEAN;
    impure function detected return BOOLEAN;
    impure function fault_name return STRING;
    procedure set_simulate;
    procedure clr_simulate;
    procedure set_detected;
  end protected fault_list;
  shared variable fault_sim : fault_list;
end package fault_inject;
```

The 10 function and procedure definitions are now placed within the **protected body** of the type, which in turn is placed in a **package body**. Note that the data structure type itself, together with local variables, is also declared here.

```vhdl
use STD.textio.all; -- contains definition of line

package body fault_inject is
  type fault_list is protected body
    type fault_model; -- incomplete type declaration
    type fault_ptr is access fault_model;

    type fault_model is
      record
        fault_name : line; -- line is access string
        simulating : BOOLEAN;
        detected : BOOLEAN;
        index : NATURAL; -- to allow unique reference
        next_fault : fault_ptr;
      end record fault_model;

    variable fault_head, present_fault: fault_ptr := null;
    variable fault_count : NATURAL := 0;
```

```
impure function new_fault(name : STRING) return
  NATURAL is
begin
  fault_count := fault_count + 1;
  fault_head := new fault_model'(new STRING'(name),
      FALSE, FALSE
      fault_count,
      fault_head);
  return (fault_count);
end function new_fault;

procedure first_fault is
begin
  present_fault := fault_head;
end procedure first_fault;

impure function end_fault_list return BOOLEAN is
begin
  return present_fault = null;
end function end_fault_list;

procedure inc_fault_list is
begin
  if present_fault /= null then
    present_fault := present_fault.next_fault;
  end if;
end procedure inc_fault_list;

impure function simulating(fault_no : NATURAL)
      return BOOLEAN is
begin
  if present_fault /= null then
    return present_fault.index = fault_no and
      present_fault.simulating;
  else
    return FALSE;
  end if;
end function simulating;

impure function detected return BOOLEAN is
begin
  if present_fault /= null then
    return present_fault.detected;
  else
    return FALSE;
  end if;
end function detected;
```

```vhdl
    impure function fault_name return STRING is
    begin
      if present_fault /= null then
        return present_fault.fault_name.all;
      else
        return "";
      end if;
    end function fault_name;

    procedure set_simulate is
    begin
      if present_fault /= null then
        present_fault.simulating : = TRUE;
      end if;
    end procedure set_simulate;

    procedure clr_simulate is
    begin
      if present_fault /= null then
        present_fault.simulating : = FALSE;
      end if;
    end procedure clr_simulate;

    procedure set_detected is
    begin
      if present_fault /= null then
        present_fault.detected : = TRUE;
      end if;
    end procedure set_detected;

  end protected body fault_list;
end package body fault_inject;
```

At the time of writing, the 2002 VHDL standard had not appeared in any simulator. Therefore the package definition has to be changed to be compliant with the 1993 standard. The following code 'fakes' the 2002-compliant package.

```vhdl
package fault_sim is
    impure function new_fault(name : STRING)
      return NATURAL;
    procedure first_fault;
    impure function end_fault_list return BOOLEAN;
    procedure inc_fault_list;
    impure function simulating(fault_no : NATURAL)
      return BOOLEAN;
    impure function detected return BOOLEAN;
    impure function fault_name return STRING;
```

```
    procedure set_simulate;
    procedure clr_simulate;
    procedure set_detected;
end package fault_sim;
```

Notice that the package name has been changed to that of the protected type. In order to keep the **use** clauses the same, we now need the following package:

```
package fault_inject is
  alias fault_sim is WORK.fault_sim;
end package fault_inject;
```

Finally, the package body for fault_sim can be defined:

```
use STD.textio.all; -- contains definition of line
package body fault_sim is
  type fault_model; -- incomplete type declaration
  type fault_ptr is access fault_model;
  type fault_model is
    record
      fault_name : line; -- line is access string
      simulating : BOOLEAN;
      detected : BOOLEAN;
      index : NATURAL; --added to allow unique reference
      next_fault : fault_ptr;
    end record fault_model;

  shared variable fault_head, present_fault: fault_ptr
    := null;
  shared variable fault_count : NATURAL := 0;
-- The procedure and function definitions are identical
-- to those in the 2002 version.
end package body fault_sim;
```

In Chapter 12, a random number generator that used a 2002 compliant shared variable was shown. The following emulates that package, but is compliant with the 1993 standard. Here, the shared variable is hidden in the package body, so the emulation is much simpler.

```
package random is
  impure function rand return REAL;
  impure function negexp(t : TIME) return TIME;
end package random;

library IEEE;
use IEEE.math_real.all;
package body random is
  type seed_pair is record
    seed1, seed2 : INTEGER;
  end record seed_pair;
```

```
   impure function init_seed return seed_pair is
-- same as in Chapter 12

   end function init_seed;

   shared variable vseed : seed_pair := init_seed;

   impure function rand return REAL is
     variable seeds : seed_pair;
     variable rnd : REAL;
   begin
     seeds := vseed;
     uniform (seeds.seed1, seeds.seed2, rnd);
     vseed := seeds;
     return rnd;
   end function rand;

   function negexp(t : TIME) return TIME is
   begin
     return INTEGER(-log(rand)*(REAL(t / NS))) * NS;
   end function negexp;

end package body random;
```

For both the 1993 and 2002 compliant versions, the package is referenced by including the clause:

```
use WORK.random.all;
```

Bibliography

[1] Abramovici, M., Breuer, M.A. and Friedman, A.D. (1990). *Digital System Testing and Testable Design*, revised printing. IEEE Press, New York.

[2] Aftabjahani, S.A. and Navabi, N. (1997). 'Functional fault simulation of VHDL gate level models'. *Proceedings VHDL International Users' Forum*, pp. 18–23, October.

[3] Ashenden, P.J. (1998). 'SUAVE: VHDL extensions for system-level modeling'. *VHDL International Users' Forum*, October.

[4] Bergeron, J. (2003). *Writing Testbenches: Functional Verification of HDL Models*, 2nd edn. Kluwer Academic Publishers, Boston, MA.

[5] De Micheli, G. (1994). *Synthesis and Optimization of Digital Circuits*. McGraw-Hill, New York.

[6] Dewey, A. (1997). *Analysis and Design of Digital Systems with VHDL*. PWS Publishing Company, Boston, MA.

[7] Edwards, M.D. (1992). *Automatic Logic Synthesis Techniques for Digital Systems*. Macmillan, Basingstoke.

[8] Hamming, R.W. (1980). *Coding and Information Theory*. Prentice-Hall, Englewood Cliffs, NJ.

[9] Hennessy, J.L. and Patterson, D.A. (1990). *Computer Architecture: a Quantitative Approach*. Morgan Kaufmann, San Francisco.

[10] Hewlett-Packard (1990). *HP Boundary-Scan Tutorial and BSDL Reference Guide*. Hewlett-Packard Company, Palo Alto, CA.

[11] Hill, F.J. and Peterson, G.R. (1993). *Computer Aided Logical Design with Emphasis on VLSI*, 4th edn. John Wiley & Sons, New York.

[12] Horowitz, P. and Hill, W. (1989). *The Art of Electronics*. Cambridge University Press, New York.

[13] Litovski, V. and Zwolinski, M. (1997). *VLSI Circuit Simulation and Optimization*. Chapman & Hall, London.

[14] Maccabe, A.B. (1993). *Computer Systems: Architecture, Organization and Programming*. Richard D. Irwin, Homewood, IL.

[15] Maunder, C. (1992). *The Board Designer's Guide to Testable Logic Circuits*. Addison-Wesley, Reading, MA.

[16] Miczo, A. (1987). *Digital Logic Testing and Simulation*. John Wiley & Sons, New York.

[17] Molenkamp, E. and Mekenkamp, G. (1997). 'Processes with "incomplete" sensitivity lists and their synthesis aspects'. *Proceedings VHDL International Users' Forum*, pp. 75–81, October.

[18] Morison, J.D. and Clarke, A.S. (1994) *ELLA 2000*. McGraw-Hill, New York.

[19] Navabi, Z. (1993). *VHDL Analysis and Modeling of Digital Systems*. McGraw-Hill, New York.

[20] Nixon, M.S. (1995). *Introductory Digital Design: a Programmable Approach*. Macmillan, Basingstoke.

[21] Perry, D.L. (1994). *VHDL*, 2nd edn. McGraw-Hill, New York.

[22] Rushton, A. (1998). *VHDL for Logic Synthesis*, 2nd edn. John Wiley & Sons, New York.

[23] Skahill, K. (1996). *VHDL for Programmable Logic*. Addison-Wesley, Reading, MA.

[24] Smith, D.J. (1996). *HDL Chip Design*. Doone Publishing, Austin, TX.

[25] Unger, S.H. (1995). 'Hazards, critical races, and metastability'. *IEEE Transactions on Computers*, **44**(6), 754–68.

[26] Wakerley, J.F. (2002). *Digital Design Principles and Practices*, 3rd edn. Prentice Hall, Englewood Cliffs, NJ.

[27] Weste, N.H.E. and Eshraghian, K. (1992). *Principles of CMOS VLSI Design: a Systems Perspective*, 2nd edn. Addison-Wesley, Reading, MA.

[28] Weyerer, M. and Goldemund, G. (1992). *Testability of Electronic Circuits*. Carl Hanser Verlag, Munich and Vienna, and Prentice Hall International, Englewood Cliffs, NJ.

[29] Wilkins, B.R. (1986). *Testing Digital Circuits*. Van Nostrand Reinhold (UK), Wokingham.

[30] Wilkinson, B. (1992). *Digital System Design*, 2nd edn. Prentice Hall, Englewood Cliffs, NJ.

[31] Wolf, W. (1994). *Modern VLSI Design: a Systems Approach*. Prentice Hall, Englewood Cliffs, NJ.

Answers to selected exercises

....

3.3

```
entity Nand3 is
  port (w, x, y : in BIT; z: out BIT);
end entity Nand3;

architecture ex1 of Nand3 is
begin
  z <= not (w and x and y) after 5 NS;
end architecture ex1;
```

3.5

```
entity FullAdder is
  port (a, b, Ci : in BIT; S, Co: out BIT);
end entity FullAdder;

architecture netlist of FullAdder is
  signal na, nb, nc, d, e, f, g, h, i, j : BIT;
begin
  n0 : entity WORK.Not1 port map (a, na);
  n1 : entity WORK.Not1 port map (b, nb);
  n2 : entity WORK.Not1 port map (Ci, nc);
  a0 : entity WORK.And3 port map (na, nb, Ci, d);
  a1 : entity WORK.And3 port map (na, b, nc, e);
  a2 : entity WORK.And3 port map (a, b, Ci, f);
  a3 : entity WORK.And3 port map (a, nb, nc, g);
  o0 : entity WORK.Or4 port map (d, e, f, g, S);
```

```
   a4 : entity WORK.And2 port map (b, Ci, h);
   a5 : entity WORK.And2 port map (a, b, i);
   a6 : entity WORK.And2 port map (a, Ci, j);
   o1 : entity WORK.Or3 port map (h, i, j, Co);
end architecture netlist;
```

3.6

```
entity testadder is
end entity testadder;


architecture ta of testadder is
  signal a, b, Ci, S, Co : BIT;
begin
  f0 : entity WORK.FullAdder port map(a, b, Ci, S, Co);
  a <= '0', '1' after 10 NS, '0' after 20 NS,
       '1' after 30 NS, '0' after 40 NS, '1' after 50 NS,
       '0' after 60 NS, '1' after 70 NS;
  b <= '0', '1' after 20 NS, '0' after 40 NS,
       '1' after 60 NS;
  Ci <= '0', '1' after 40 NS;
end architecture ta;
```

4.3

```
library IEEE;
use IEEE.std_logic_1164.all;


entity decoder is
  port (a : in std_logic_vector(2 downto 0);
        z : out std_logic_vector(7 downto 0));
end entity decoder;


architecture bool_expr of decoder is
begin
  z(0) <= not a(0) and not a(1) and not a(2);
  z(1) <= a(0) and not a(1) and not a(2);
  z(2) <= not a(0) and a(1) and not a(2);
  z(3) <= a(0) and a(1) and not a(2);
  z(4) <= not a(0) and not a(1) and a(2);
  z(5) <= a(0) and not a(1) and a(2);
```

```vhdl
    z(6) <= not a(0) and a(1) and a(2);
    z(7) <= a(0) and a(1) and a(2);
end architecture bool_expr;

architecture when_else of decoder is
begin
    z <= "00000001" when a = "000" else
         "00000010" when a = "001" else
         "00000100" when a = "010" else
         "00001000" when a = "011" else
         "00010000" when a = "100" else
         "00100000" when a = "101" else
         "01000000" when a = "110" else
         "10000000" when a = "111" else
         "XXXXXXXX";
end architecture when_else;

architecture with_select of decoder is
begin
    with a select
      z <= "00000001" when "000",
           "00000010" when "001",
           "00000100" when "010",
           "00001000" when "011",
           "00010000" when "100",
           "00100000" when "101",
           "01000000" when "110",
           "10000000" when "111",
           "XXXXXXXX" when others;
end architecture with_select;

library IEEE;
use IEEE.std_logic_1164.all;

entity testdecode is
end entity testdecode;

architecture td of testdecode is
  signal a : std_logic_vector(2 downto 0);
  signal z0, z1, z2 : std_logic_vector(7 downto 0);
begin
  d0 : entity WORK.decoder(bool_expr) port map (a, z0);
  d1 : entity WORK.decoder(when_else) port map (a, z1);
  d2 : entity WORK.decoder(with_select) port map (a, z2);
  a <= "000", "001" after 10 NS, "010" after 20 NS;
end architecture td;
```

4.4

```vhdl
library IEEE;
use IEEE.std_logic_1164.all, IEEE.numeric_std.all;

entity priority is
  generic (n : positive);
  port (a: in std_logic_vector(2**n-1 downto 0);
        y: out std_logic_vector(n-1 downto 0);
        valid: out std_logic);
end entity priority;

architecture iterative of priority is
begin
  process (a) is
  begin
    valid <= '0';
    y <= (others => '0');
    for i in a'RANGE loop
      if a(i) = '1' then
        y <= std_logic_vector(to_unsigned(i, n));
        valid <= '1';
        exit;
      end if;
    end loop;
  end process;
end architecture iterative;
```

4.5

```vhdl
library IEEE;
use IEEE.std_logic_1164.all;

entity comparator is
  port (x, y : in std_logic_vector;
        eq : out std_logic);
end entity comparator;

architecture iterative of comparator is
begin
  process (x,y) is
    variable eqi : std_logic;
  begin
    eqi := '1';
    for i in x'range loop
      eqi := eqi and (x(i) xnor y(i));
    end loop;
```

```vhdl
      eq <= eqi;
  end process;
end architecture iterative;
```

4.6

```vhdl
package quad_logic is
  type quad is ('0', '1', 'Z', 'X');
  type quad_vector is array (natural range <>) of quad;
  function "and" (Left, Right: quad) return quad;
  function resolved (arg : quad_vector) return quad;
  subtype quad_wire is resolved quad;
end package quad_logic;

package body quad_logic is
  function "and" (Left, Right: quad) return quad is
  type quad_array is array (quad, quad) of quad;
  constant and_table : quad_array := (('0', '0', '0', '0'),
                                      ('0', '1', '1', 'X'),
                                      ('0', '1', '1', 'X'),
                                      ('0', 'X', 'X', 'X'));
  begin
    return and_table(Left, Right);
  end function "and";
  function resolved (arg : quad_vector) return quad is
    variable temp : quad := '1';
  begin
    for i in arg'RANGE loop
      temp := temp and arg(i);
    end loop;
    return temp;
  end function resolved;
end package body quad_logic;


use work.quad_logic.all;


entity nand2 is
  port (a, b : in quad; z : out quad);
end entity nand2;

architecture open_drain of nand2 is
begin
  z <= '0' when a = '1' and b = '1' else
       'Z' when a = '0' or b = '0' else
       'X';
end architecture open_drain;
```

```vhdl
use work.quad_logic.all;

entity testbus is
end entity testbus;

architecture tb of testbus is
  signal a, b : quad_vector (3 downto 0);
  signal q : quad_wire;
begin
  n0 : entity WORK.nand2 port map (a(0), b(0), q);
  n1 : entity WORK.nand2 port map (a(1), b(1), q);
  n2 : entity WORK.nand2 port map (a(2), b(2), q);
  n3 : entity WORK.nand2 port map (a(3), b(3), q);
  a <= "1111";
  b <= "0000", "0001" after 10 NS, "0000" after 20 NS,
       "0010" after 30 NS;
end architecture tb;
```

5.3

```vhdl
seq: process (clock, reset) is
begin
  if reset = '1' then
    state <= S0;
  elsif rising_edge(clock) then
    state <= next_state;
  end if;
end process seq;
```

5.5

```vhdl
entity state_machine is
  port(x, clock, reset : in BIT;
       z : out BIT);
end entity state_machine;

architecture behaviour of state_machine is
  type state_type is (S0, S1, S2);
  signal state, next_state : state_type;
begin
  seq: process (clock, reset) is
  begin
    if reset = '1' then
      state <= S0;
    elsif clock = '1' and clock'EVENT then
      state <= next_state;
```

```vhdl
      end if;
    end process seq;
    com: process (state, x) is
    begin
      Z <= '0';
      case state is
        when S0 =>
          if X = '0' then
            next_state <= S0;
          else
            next_state <= S1;
          end if;
        when S1 =>
          if X = '0' then
            next_state <= S0;
          else
            next_state <= S2;
          end if;
        when S2 =>
          if X = '0' then
            next_state <= S0;
          else
            Z <= '1';
            next_state <= S2;
          end if;
      end case;
    end process com;
end architecture behaviour;
```

5.8

```vhdl
entity state_machine is
  port(a, b, clock, reset : in BIT;
       z : out BIT);
end entity state_machine;

architecture one of state_machine is
  type state_type is (S0, S1, S2);
  signal state, next_state : state_type;
begin
  seq: process (clock, reset) is
  begin
    if reset = '1' then
      state <= S0;
    elsif clock = '1' and clock'EVENT then
      state <= next_state;
```

```vhdl
    end if;
  end process seq;
com: process (state, a, b) is
begin
  Z <= '0';
  case state is
    when S0 =>
      if a = '1' and b = '1' then
        next_state <= S1;
      else
        next_state <= S0;
      end if;
    when S1 =>
      if a = '1' and b = '0' then
        next_state <= S2;
      else
        next_state <= S0;
      end if;
    when S2 =>
      if a = '0' and b = '0' then
        Z <= '1';
      end if;
      next_state <= S0;
  end case;
  end process com;
end architecture one;
```

5.9

```vhdl
architecture three of state_machine is
  type state_type is (S0, S1, S2);
  signal state, next_state : state_type;
begin
  seq: process (clock, reset) is
  begin
    if reset = '1' then
      state <= S0;
    elsif clock = '1' and clock'EVENT then
      state <= next_state;
    end if;
  end process seq;
  com: process (state, a, b) is
  begin
    case state is
      when S0 =>
```

```vhdl
      if a = '1' and b = '1' then
        next_state <= S1;
      else
        next_state <= S0;
      end if;
    when S1 =>
      if a = '1' and b = '0' then
        next_state <= S2;
      else
        next_state <= S0;
      end if;
    when S2 =>
      next_state <= S0;
    end case;
  end process com;
  Z <= '1' when state = S2 and a = '0' and b = '0' else '0';
end architecture three;
```

5.10

```vhdl
architecture one of state_machine is
begin
  seq: process (clock, reset) is
    type state_type is (S0, S1, S2);
    variable state : state_type;
  begin
    Z <= '0';
    if reset = '1' then
      state := S0;
    elsif clock = '1' and clock'EVENT then
      case state is
        when S0 =>
          if a = '1' and b = '1' then
            state := S1;
          else
            state := S0;
          end if;
        when S1 =>
          if a = '1' and b = '0' then
            state := S2;
          else
            state := S0;
          end if;
        when S2 =>
          if a = '0' and b = '0' then
```

```
                    Z <= '1';
                end if;
                state := S0;
            end case;
        end if;
    end process seq;
end architecture one;
```

6.3

```vhdl
library IEEE;
use IEEE.std_logic_1164.all;

entity D_FF is
    generic (Setup, Hold: TIME := 3 NS);
    port (D, Clk, Set, Reset: in std_logic;
          Q : out std_logic);
begin
    assert (not(Clk = '1' and Clk'EVENT and not
            D'STABLE(Setup)))
    report "Setup time violation" severity WARNING;
    assert (not(Clk = '1' and D'EVENT and not
            Clk'STABLE(Hold)))
    report "Hold time violation" severity WARNING;
end entity D_FF;

architecture behavioural of D_FF is
begin
    p0 : process (Clk, Set, Reset) is
    begin
        if Set = '0' then
            Q <= '1';
        elsif Reset = '0' then
            Q <= '0';
        elsif rising_edge(Clk) then
            Q <= D;
        end if;
    end process p0;
end architecture behavioural;
```

6.6

```vhdl
library IEEE;
use IEEE.std_logic_1164.all;
use IEEE.numeric_std.all;
```

```vhdl
entity Counter is
  generic(N : POSITIVE := 8);
  port(Clk, Reset, Up : in std_logic;
       Count : out std_logic_vector(N-1 downto 0));
end entity Counter;

architecture Rtl of Counter is
  signal Cnt : unsigned(N-1 downto 0);
  constant Cmax : unsigned(N-1 downto 0) := (others => '1');
  constant Cmin : unsigned(N-1 downto 0) := (others => '0');
begin
  process(Clk, Reset) is
  begin
    if Reset = '1' then
      Cnt <= (others => '0');
    elsif rising_edge(Clk) then
      if Up = '1' and Cnt < Cmax then
        Cnt <= Cnt + 1;
      elsif Up = '0' and Cnt > Cmin then
        Cnt <= Cnt - 1;
      end if;
    end if;
  end process;
  Count <= std_logic_vector(Cnt);
end architecture Rtl;
```

6.7

```vhdl
library IEEE;
use IEEE.std_logic_1164.all;

entity piso is
  generic(n : NATURAL := 8);
  port(a : in std_logic_vector(n-1 downto 0);
       q : out std_logic;
       clk, load : in std_logic);
end entity piso;

architecture rtl of piso is
begin
  p0: process (clk) is
    variable reg : std_logic_vector(n-1 downto 0);
  begin
    if rising_edge(clk) then
      if load = '1' then
        reg := a;
```

```
      else
        reg := ('0' & reg(n-1 downto 1));
      end if;
      q <= reg(0);
    end if;
  end process p0;
end architecture rtl;
```

6.11

```
library IEEE;
use IEEE.std_logic_1164.all;

entity counter is
  port(clk : in std_logic;
       reset : in std_logic;
       count : out std_logic_vector(2 downto 0));
end entity counter;

architecture lfsr of counter is
begin
  p0: process (clk, reset) is
    variable reg : std_logic_vector(2 downto 0);
  begin
    if reset = '1' then
      reg := (others => '1');
    elsif rising_edge(clk) then
      reg := reg(1 downto 0) & (reg(2) xor reg(1));
    end if;
    count <= reg;
  end process p0;
end architecture lfsr;
```

7.6

State s9 is modified to load the PC from the Addr part of the IR.

```
when s9 =>
  Addr_bus <= '1';
  load_PC <= '1';
  next_state <= s0;
```

In fact, this could be done in the same clock cycle as s6; thus s9 becomes a conditional output for s6:

```
when s6 =>
  CS <= '1';
```

```
    R_NW <= '1';
    if op = load then
      next_state <= s7;
    elsif op = bne then
      if z_flag = '0' then
        Addr_bus <= '1';
        load_PC <= '1';
      end if
      next_state <= s0;
    else
      next_state <= s8;
    end if;
```

8.4

```
architecture correct1 of mux is
  signal sel : integer range 0 to 1;
begin
  m1: process is
  begin
    sel <= 0;
    wait for 0 ns;
    if (c = '1') then
      sel <= sel + 1;
    end if;
    wait for 0 ns;
    case sel is
      when 0 =>
        z <= a;
      when 1 =>
        z <= b;
    end case;
    wait on a, b, c;
  end process m1;
end architecture correct1;

architecture correct2 of mux is
begin
  m1: process (a, b, c) is
    variable sel : integer range 0 to 1;
  begin
    sel := 0;
    if (c = '1') then
      sel := sel + 1;
    end if;
    case sel is
```

```
      when 0 =>
        z <= a;
      when 1 =>
        z <= b;
    end case;
  end process m1;
end architecture correct2;
```

9.2

```
library IEEE;
use IEEE.std_logic_1164.all, IEEE.numeric_std.all;

entity counter is
  generic(n : NATURAL := 4);
  port(clk : in std_logic;
       reset : in std_logic;
       ready : out std_logic);
end entity counter;

architecture fsm of counter is
begin
  p0: process (clk, reset) is
    attribute enum_encoding : string;
    type state_type is (s0, s1, s2, s3, s4, s5, s6, s7);
    attribute enum_encoding of state_type: type is
        "0000 0001 0011 0111 1111 1110 1100 1000";
    variable state : state_type;
  begin
    if reset = '1' then
      state := s0;
      ready <= '1';
    elsif rising_edge(clk) then
      ready <= '0';
      case state is
        when s0 =>
          state := s1;
          ready <= '1';
        when s1 => state := s2;
        when s2 => state := s3;
        when s3 => state := s4;
        when s4 => state := s5;
        when s5 => state := s6;
        when s6 => state := s7;
        when s7 => state := s0;
      end case;
    end if;
```

```vhdl
    end process p0;
  end architecture fsm;


9.4

library IEEE;
use IEEE.std_logic_1164.all;

entity fsm is
  port (clk, a, reset: in std_logic;
        y: out std_logic);
end entity fsm;

architecture try1 of fsm is
  type statetype is (s0, s1, s2);
  signal currentstate, nextstate : statetype;
begin
  seq: process (clock, reset) is
  begin
    if reset = '1' then
      currentstate <= s0;
    elsif rising_edge(clock) then
      currentstate <= nextstate;
    end if;
  end process seq;
  com: process (currentstate, a) is
  begin
    y <= '0';
    case currentstate is
      when s0 =>
        if a = '1' then
          nextstate <= s1;
        else
          nextstate <= s2;
        end if;
      when s1 =>
        y <= '1';
        nextstate <= s0;
      when s2 =>
        if a = '1' then
          nextstate <= s2;
        else
          nextstate <= s0;
        end if;
    end case;
  end process com;
end architecture try1;
```

10.3

Test for $A/0$: 0100/0; also covers $E/1$, $G/1$, $H/0$, $I/0$ and $J/1$.

Test for $A/1$: 1100/1; also covers $B/0$, $C/1$, $D/1$, $E/0$, $F/0$, $H/1$ and $J/0$.

Test for $G/0$ implies $G = 1$, hence $B = C = 1$. To propagate G to I implies $F = 1$, which implies $C = D = 0$. Hence there is a contradiction.

10.7

$11 \ldots 11/0, 11 \ldots 10/1, 11 \ldots 01/1, \ldots, 10 \ldots 11/1, 01 \ldots 11/1$.

11.3

50 flip-flops implies $2^{50} \simeq 10^{15}$ states. At 1 MHz it would take 10^9 seconds $\simeq 32$ years to reach all states.

It takes 50 clock cycles to load the scan path (unloading can be done at the same time as loading the next pattern). 200 patterns take 10 000 cycles = 10 ms at 1 MHz.

11.7

State	TMS	TDI
Test-Logic-Reset	0	–
Run-Test/Idle	1	–
Select-DR-Scan	1	–
Select-IR-Scan	0	–
Capture-IR	0	–
Shift-IR	0	0
Shift-IR	1	1
Exit1-IR	1	–
Update-IR	1	–
Select-DR-Scan	0	–
Capture-DR	0	–
Shift-DR	0	0
Shift-DR	0	1
Shift-DR	0	0
Shift-DR	1	1
Exit1-DR	1	–
Update-DR	0	–
Run-Test/Idle		

– means 'don't care'. Change of state occurs on rising edge of TCK.

11.10

```vhdl
entity tap_controller is
  port (tms, tck : in BIT;
        ShiftIR, ClockIR, UpdateIR,
        ShiftDR, ClockDR, UpdateDR : out BIT);
end entity tap_controller;

architecture fsm of tap_controller is
  type state is (test_logic_reset, run_test_idle,
                 select_DR_scan, capture_DR,
                 shift_DR, exit1_DR, pause_DR,
                 exit2_DR, update_DR, select_IR_scan,
                 capture_IR, shift_IR, exit1_IR,
                 pause_IR, exit2_IR, update_IR);
  signal current_state, next_state : state;
begin
  seq: process is
  begin
    wait until tck = '1';
    current_state <= next_state;
  end process seq;
  com: process (tms, current_state) is
  begin
    ShiftIR <= '0';
    ClockIR <= '0';
    UpdateIR <= '0';
    ShiftDR <= '0';
    ClockDR <= '0';
    UpdateDR <= '0';
    case current_state is
      when test_logic_reset =>
        if tms = '0' then
          next_state <= run_test_idle;
        else
          next_state <= test_logic_reset;
        end if;
      when run_test_idle =>
        if tms = '1' then
          next_state <= select_DR_scan;
        else
          next_state <= run_test_idle;
        end if;
      when select_DR_scan =>
        if tms = '1' then
          next_state <= select_IR_scan;
```

```
      else
        next_state <= capture_DR;
      end if;
    when capture_DR =>
      ClockDR <= '1';
      if tms = '1' then
        next_state <= exit1_DR;
      else
        next_state <= shift_DR;
      end if;
    when shift_DR =>
      ClockDR <= '1';
      ShiftDR <= '1';
      if tms = '1' then
        next_state <= exit1_DR;
      else
        next_state <= shift_DR;
      end if;
    when exit1_DR =>
      if tms = '1' then
        next_state <= update_DR;
      else
        next_state <= pause_DR;
      end if;
    when pause_DR =>
      if tms = '1' then
        next_state <= exit2_DR;
      else
        next_state <= pause_DR;
      end if;
    when exit2_DR =>
      if tms = '1' then
        next_state <= update_DR;
      else
        next_state <= shift_DR;
      end if;
    when update_DR =>
      UpdateDR <= '1';
      if tms = '1' then
        next_state <= select_DR_scan;
      else
        next_state <= run_test_idle;
      end if;
    when select_IR_scan =>
      if tms = '1' then
        next_state <= test_logic_reset;
```

```vhdl
          else
            next_state <= capture_IR;
          end if;
        when capture_IR =>
          ClockIR <= '1';
          if tms = '1' then
            next_state <= exit1_IR;
          else
            next_state <= shift_IR;
          end if;
        when shift_IR =>
          ClockIR <= '1';
          ShiftIR <= '1';
          if tms = '1' then
            next_state <= exit1_IR;
          else
            next_state <= shift_IR;
          end if;
        when exit1_IR =>
          if tms = '1' then
            next_state <= update_IR;
          else
            next_state <= pause_IR;
          end if;
        when pause_IR =>
          if tms = '1' then
            next_state <= exit2_IR;
          else
            next_state <= pause_IR;
          end if;
        when exit2_IR =>
          if tms = '1' then
            next_state <= update_IR;
          else
            next_state <= shift_IR;
          end if;
        when update_IR =>
          UpdateIR <= '1';
          if tms = '1' then
            next_state <= select_IR_scan;
          else
            next_state <= run_test_idle;
          end if;
      end case;
  end process com;
end architecture fsm;
```

12.6

States A, E and F can be merged. States B and C can be merged. An extra state (T) needs to be introduced to avoid races – let this be between D and AEF. A possible state assignment is AEF (00), BC (01), D (11), T (10), giving next state and output equations:

$$Y_1{}^+ = Y_1.Y_0 + \overline{P}.R.Y_0$$
$$Y_0{}^+ = P.\overline{R} + \overline{P}.R.Y_0 + \overline{P}.\overline{Y_1}.Y_0$$
$$Q = Y_1$$

12.7

There are three feedback loops in Figure 12.4. Insert a virtual buffer at A (Y_1), between F and the NAND gate with output B (Y_2) and at Q (Y_3).

$$Y_1{}^+ = D.R.Y_2 + \overline{S} + Y_1.R.C$$
$$Y_2{}^+ = Y_1.R + \overline{C} + Y_2.D.R$$
$$Y_3{}^+ = Y_3.Y_1.R + Y_3.R.\overline{C} + Y_3.Y_2.D.R + Y_1.R.C + \overline{S}$$

13.1

```
library IEEE;
use IEEE.electrical_systems.all;

entity inductor is
  generic (L: REAL);
  port (terminal node1, node2: electrical);
end entity inductor;

architecture didt of inductor is
  quantity vl across il through node1 to node2;
begin
  vl == L*il'DOT;
end architecture didt;
```

13.3

```
library IEEE;
use IEEE.electrical_systems.all;

entity vramp is
  generic (vl, vh, td, tr: REAL);
  port (terminal node1, node2: electrical);
end entity vramp;

architecture sim of vramp is
  quantity vr across ir through node1 to node2;
```

```
begin
  if NOW < td use
    vr == vl;
  elsif NOW < td+tr use
    vr == vl + (NOW - td)*(vh - vl)/tr;
  else
    vr == vh;
  end use;
end architecture sim;
```

Index